jfUNUレクチャー・シリーズ 10

持続可能な地球社会をめざして

わたしの SDGs への取組み

勝間 靖 編

国際書院

UNU Alumni Association 10th Anniversary
UNU/jfUNU Junior Fellows Symposium 2017

Towards Sustainable Global Society:
My Challenge for the SDGs
by
Yasushi Katsuma ed.
Copyright ©2017 by Japan Foundation
for United Nations University (jfUNU)
ISBN978-4-87791-292-5 C3032 ¥2000E Printed in Japan

目　　次

UNU Alumni Association 創立 10 周年記念
UNU/jfUNU ジュニアフェローシンポジウム 2017

持続可能な地球社会をめざして：
わたしの SDGs への取組み

目　次

はじめに……………………………………沖　大幹　7
シンポジウムの意義………………………勝間　靖　11

第 1 部　基調講演

持続可能な開発目標（SDGs）達成に向けた国連と日本の役割
………………………………………………弓削昭子　17

第 2 部　持続可能な地球社会をめざして

1　持続可能な社会とグローバル・ガバナンス
　　………………………………………滝沢美佐子　63
2　地球環境の変化とレジリエンスの観点から
　　………………………………………吉高神　明　73
3　持続可能な開発目標（SDGs）とグローバル・シティズンシップ：
　持続可能な社会を支える人々の協力・協働・共生
　　………………………………………杉村美紀　81

4 持続可能な開発目標（SDGs）達成に向けた
  国連大学の取組みと今後の展望……………齊藤　修　89
5 持続可能な開発目標（SDGs）とビジネス：
  グローバル規範構築の可能性………………秋月弘子　97

第3部　「誰も置き去りにしない」社会をめざして

「誰も置き去りにしない」：
持続可能なグローバル社会のためのガバナンス…勝間　靖　107

第4部　　わたしのSDGsへの取組み

セッション1：持続可能な社会とグローバル・ガバナンス
……………………………………………………………132
セッション2：地球環境の変化とレジリエンス……………160
セッション3：グローバル・シティズンシップ……………178

あとがき………………………………………森　茜　197

付録
持続可能な開発目標（SDGs）……………………………201

執筆者・編者紹介……………………………………………205
索引……………………………………………………………211

はじめに

　国連大学は、1969 年、当時のウ・タント国連事務総長による「真に国際的な性格を有し、国連憲章が定める平和と進歩の諸目的に合致する国際連合の大学」の設立提案に始まります。当時の日本の総理大臣佐藤栄作氏がその構想に共感し、翌年 1970 年には日本政府が国連大学設立へのサポートを表明しました。1972 年の国連総会で国連大学の設立が承認され、1973 年の国連総会で国連大学の設置についての約款ともいうべき「国連大学憲章」が採択されました。

　こうして 1975 年、国連大学は、日本（東京）に本部を置く唯一の国連機関として活動を開始しました。以来、45 年にわたって、国連のシンクタンクとして、人類の存続・発展・福祉にかかわる緊急かつ世界的・地球規模的な課題の解決に関係した研究、人材育成、および知識の普及といった活動で実績を積んでいます。

　現在では、世界十数ケ国に研究拠点を持ち、世界各国の研究機関等との協働により、国連やその加盟国が直面する喫緊の課題の解決へ向けて、調査分析に基づく信頼性の高い助言や提言を提供しています。

　創設後の 35 年間、国連大学は固有の大学院機能を持たず、数日から 1 か月程度の短期人材育成コース（キャパシティ・デベロップメント・コース）を開催していました。しかしながら、

学位プログラムに対して盛り上がる期待に応えて 2010 年には国連総会で国連大学憲章が改正され、国連大学傘下のいくつかの研究所に大学院修士課程と博士課程が設置されました。現在では名実ともに大学院大学として、地球の持続可能性と人類の発展にかかわる実践機関や研究機関に有為な人材を送り出しています。

国連大学のキャパシティ・デベロップメント・コースや大学院の修了生たちは、"UNU Alumni Association（UNU 同窓会）"を組織し、たえざる連携と友好を図っています。初期の修了生たちの中には大学教授になった同窓生や国連機関の中枢で仕事をしている同窓生もいます。

本書は、その"UNU Alumni Association（UNU 同窓会）"の創立 10 周年を記念して開催された UNU/jfUNU ジュニアフェローシンポジウムの成果として、修了生たちが世界の現場でどのような課題に直面し、どのような活動をしているのかを持ち寄り、国連の 2030 年アジェンダが掲げている持続可能な開発目標（Sustainable Development Goals: SDGs）にかかわる多様な取り組みを議論した結果をとりまとめたものです。

SDGs については本書の中でその成立の経緯や世界各地での日本をはじめとする各国政府レベルから草の根に至る様々なレベルでの取り組みが詳細に紹介されているとおり、2015 年 9 月の「国連持続可能な開発サミット」で採択された「持続可能な開発のための 2030 アジェンダ」（2030 アジェンダ）の中核をなす 21 世紀における世界の大義名分です。

2030 アジェンダが採択された際のスローガンは"誰一人取

り残さない（no one will be left behind）"として知られていますが、このSDGsはミレニアム開発目標（Millennium Development Goals: MDGs）に比べると極めて理想主義的で「誰一人文句が言えない」野心的な目標設定となっています。2030アジェンダと同じく2015年に採択された気候変動対策の国際的な目標を定めた「パリ協定」にも現状の延長線上ではなかなか達成が困難な目標が掲げられており、2015年にはそうした野心的な主張が現実路線を凌駕する国際社会の雰囲気が醸し出されていたのかもしれません。あるいは、現実的な目標設定では取り残されてしまいそうな国や地域からの異論を抑えて文書を採択するには理想主義的な目標設定にせざるをえなかったのかもしれません。

　SDGsの究極の目標は17の目標の達成に留まらず、人類の幸福度（well-being）の増進だと考えられます。健全な自然環境、持続的な経済発展、平和で公正で包摂的な社会の三者がお互いを支えることによって自然資本、人工資本、社会資本、人的資本を形成し、水・エネルギー・食料といった基本的人権的側面を持つ財や教育・真っ当な仕事などが提供され、幸福度に直結する健康や安全、生産と消費などの持続的な開発が実現されるのではないでしょうか。SDGsとしては明記されていませんが、2030アジェンダ本文には書きこまれている重要なキーワードとして尊厳（dignity）や自由（freedom）も挙げられます。

　一方で、SDGsさえ達成できれば人類の幸福（well-being）の増進は申し分ない、というわけでもないのではないでしょうか。現状のSDGsは物質的、現世的な御利益の追求に重点が置

かれていて、精神的な豊かさや心の安寧などが目標として明確には掲げられていません。ターゲット 4.7 に文化的多様性への理解の教育が触れられているだけです。しかし、知的好奇心の充足や多様な歴史文化・精神世界の探求、あるいは芸術や娯楽コンテンツやスポーツも人類の幸福度の増進にとってやはり必須だと思います。SDGs がすべてをカバーしているわけではない点にも目配りが必要でしょう。

そういう意味では、2030 アジェンダの様な国際的な合意文書に対して、その内容を理解して従う「観客」として振る舞うばかりではなく、より良い取り組みにつながるような枠組みの提案をする「選手」として様々なステークホルダーが様々な角度から積極的に関与していくかかわりが非常に重要だと思われます。

国連大学の修了生たちは、国際社会の様々な舞台でまさにそうした「選手」として活躍できる才気を兼ね備えた人材です。本書後半には母国に限らず世界各国の現場を支えている彼ら、彼女らの報告や提言がまとめられています。SDGs をめぐるそれらの様々な思いや活動に触れて、SDGs の達成に向けた取り組みに、皆さん自身がどんなグローバルパートナーシップを構築し生かせるか、いろいろと思いをめぐらせていただけるよう祈念しています。

国連大学上級副学長、国連事務局次長補
沖　大幹

シンポジウムの意義

　2017年3月11日〜12日、東京にある国連大学（United Nations University: UNU）において、「持続可能な地球社会を目指して：私のSDGsへの取組み」と題したシンポジウムが開催されました。2015年に国連で採択された「持続可能な開発のための2030アジェンダ」と、そのために達成すべき「持続可能な開発目標（Sustainable Development Goals: SDGs）」に、私たちはいかに取り組むべきかが議論されました。

　このシンポジウムは、国連大学サステイナビリティ高等研究所（UNU Institute for the Advanced Study of Sustainability: UNU-IAS）と公益財団法人国連大学協力会（The Japan Foundation for the United Nations University: jfUNU）によって共催され、外務省と文部科学省に後援されたものです。多くの組織のパートナーシップと、多くの人びとの協力によって実現したシンポジウムでした。

　また、UNU/jfUNUジュニアフェローシンポジウムとして位置づけられ、UNUの各種の教育・研修プログラムの修了生が、かつて学んだ「母校」とも呼ぶべきUNUに集うという意味で、いわばホーム・カミングの機会だったとも言えます。2017年は、修了生によって構成されるUNU Alumni Association（UNU同窓会）が発足して10周年にあたる年でした。

　さて、本書は、このシンポジウムにおいて英語で議論された

内容の記録であると同時に、そこで触発されて、さらに考察したことを新たに執筆した原稿も含んでいます。したがって、この1年間に新たに起こったことを踏まえて、アップデートした部分もあります。

　SDGsは、すべての国連加盟国に課された課題です。もちろん、日本にとっても、日本に住むすべての人びとにとっても、重要な課題です。日本は、先進国に分類されますが、それでも、相対的貧困、生活習慣病、ジェンダー格差、エネルギー、過労死、国内の不平等、災害リスク、環境問題など、多くの問題も同時に抱えています。これらに、私たちはどのように取り組んでいくべきなのでしょうか？また、日本がこれまで多くの課題を克服・解決してきた経験を、とくに途上国へ伝えることも期待されています。私たちは、日本の国際協力にどのように貢献できるでしょうか？こうしたことも、読者の皆さんと一緒に考えていきたい点です。

　これからの数年、日本にとって、国際社会においてリーダーシップを発揮できる機会が多くあります。2019年にG20大阪サミットが開催されますが、日本が開催国となるのは初めてのことです。また、同年、第7回アフリカ開発会議が横浜で開催されます。そして、2020年には、東京オリンピック・パラリンピックが開催されます。こうした機会に、SDGsを自分たちの課題として考えることは大切でしょう。

　公益財団法人東京オリンピック・パラリンピック競技大会組織委員会（東京2020組織委員会）は、国連児童基金（United Nations Children's Fund: UNICEF）と公益財団法人日本ユニ

セフ協会と協力して、2018年の「開発と平和のためのスポーツ国際デー」（4月6日）に、「オリンピック休戦」を呼びかけるため、「折り鶴」を用いたPEACE ORIZURU（www.peaceorizuru.com）をスタートさせました。SDGsの目標16につながる、誰でも参加できる取組みの1つだと思います。

　さて、本書の出版にあたっては、多くの方にお世話になりました。まず、そもそも、UNU同窓会の発足を働きかけ、UNU/jfUNUジュニアフェローシンポジウムを企画した、jfUNUの長谷川善一専務理事及び森茜常務理事・事務局長に感謝します。また、集まった原稿の編集作業においては、小林知美さん、上田通江さん、北見瑛子さん、塩野智子さんに、ていねいな仕事をして頂きました。最後に、株式会社国際書院の石井彰社長には、本書の出版にあたって、大変にご尽力いただきました。心より感謝いたします。

　　　　　　　　　　　　　　　　　　　　編者　　勝間　靖

第 1 部　基調講演

UNU/jfUNU ジュニアフェローシンポジウム 2017
「持続可能な地球社会を目指して：私の SDGs への取組み」の
基調講演を日本語と英語で掲載しました。

持続可能な開発目標（SDGs）達成に向けた国連と日本の役割

弓 削 昭 子

　皆さま、おはようございます。ただいま、ご紹介いただきました法政大学の弓削昭子です。

　まずは、このシンポジウムで基調講演の機会をいただきましたことについて、主催者の国連大学サステイナビリティ高等研究所と公益財団法人国連大学協力会にお礼を申し上げます。

　UNU Alumni Association 創立10周年おめでとうございます。その記念として、今日のシンポジウムを開催されるのは素晴らしいことだと思います。

　今日は東日本大震災の6周年でもあります。地震、津波、原子力発電所事故などの大惨事について、そして復旧・復興、強靱な国造りについても改めて考えなければなりません。

　今日の私のお話のアウトラインは次のとおりです。まず、過去2年間の開発分野における国際社会の主要な合意を振り返ります。それらを踏まえた上で「持続可能な開発目標(Sustainable Development Goals: SDGs)」について、そしてミレニアム開発目標（Millennium Development Goals: MDGs）との違いについてお話いたします。次に持続可能な開発のための2030ア

ジェンダと SDGs の実施状況と、それに関する課題は何かを挙げさせていただきます。そして SDGs 達成に向けての国連の役割と日本の役割についてお話いたします。

1 2015-2016 年の開発分野での国際社会の合意

　過去 2 年間に開発分野で国際社会が合意した主要なものは何でしょうか。

　2015 年 3 月に仙台で開催された「第 3 回国連防災世界会議」では「仙台防災枠組 2015-2030」が採択されました。2030 アジェンダと同じ 15 年の期限つきの枠組みには、災害リスクを減らし、死者・被災者の低減、人々の健康と暮らしを守ることについての目標や取組が含まれています。

　同じ年の 7 月には「第 3 回開発資金国際会議」で「アディス・アベバ行動目標」が合意されました。持続可能な開発を進めるために、あらゆる資金調達の可能性を探ることが公約され、これには公的および民間資金、国内資金の動員および国際的な資金調達、さらに革新的な資金調達などが含まれます。

　2015 年の 9 月に国連本部で開催された国連の「持続可能な開発サミット」では、150 を超える加盟国首脳の参加のもと、その成果文書として「我々の世界を変革する：持続可能な開発のための 2030 アジェンダ」が採択されました。このアジェンダは、人間、地球および繁栄のための行動計画として、宣言と目標を掲げました。この目標が「持続可能な開発目標 (SDGs)」

です。

　同年 12 月の「国連気候変動枠組条約第 21 回締約国会議（The 21st Conference of the Parties to the United Nations Framework Convention on Climate Change: COP21）」では、気候変動への対策、適応能力の強化などを目的とする「パリ協定」が採択されました。

　2016 年、5 月の「世界人道サミット」では難民、避難民、そして人的・自然災害の被害者のため、人道支援と開発援助の強化とよりよい連携のための数多くのコミットメントが表明された上、新たなイニシアチブが立ち上がりました。

　同じ年の 9 月に開催された「難民と移民に関する国連サミット」では、難民と移民の権利を守り、人命を救うとともに、世界規模で生じている大規模な人の移動に対して責任を共有する政治的コミットメントが表明されました。

　さらに 10 月には「第 3 回国連人間居住会議（ハビタット 3）」が開催され、持続可能な都市化のための計画や管理体制などに関わる課題の解決に向けた新たな国際的な取り組み方針として「ニュー・アーバン・アジェンダ」が採択されました。

　これらのグローバルな会議とそこで採択された重要な国際合意は、すべて国連主催のもとで進められ、合意の内容は 2030 アジェンダに反映されています。

2　ミレニアム開発目標と SDGs の比較

　SDGs の前身が国際社会が合意した開発の枠組み「MDGs」

であることは皆様ご存知のとおりです。MDGsに含まれる多くの分野では重要な進展が見られました。しかし達成されなかった目標が多くあることも事実です。SDGsでは、すべてのMDGsの達成に向けての取り組みをさらに進めることが宣言されています。そして、SDGsはMDGsの枠組みをはるかに超えて、より広い範囲の課題に包括的に取り組みます。

それでは、MDGsとSDGsの違いを見てみましょう（図1参照）。

(1) MDGsは8つの目標（goals）と18のターゲットで構成されていました。

　SDGsは17の目標（goals）と169のターゲットで構成されています。

(2) MDGsの主な対象は社会開発分野でした。それに比べて、SDGsはより幅広く、包括的であり、持続可能な開発に必要な経済、社会、環境分野の目標を含んでいます。

(3) MDGsのほとんどの目標は開発途上国が対象でした。実

図1　MDGsとSDGsの比較

MDGs	SDGs
・8の目標と18のターゲット ・主な対象は社会開発分野 ・ほとんどの目標で開発途上国が対象（8つのうちの7つ） ・トップダウンのプロセスで準備作成 ・「削減」を目指す	・17の目標と169のターゲット ・幅広く、包括的。持続可能な開発に必要な経済、社会、環境分野の目標を含む ・すべての国のための普遍的な目標 ・協議を重ねる参加型プロセスで準備作成 ・「なくす」ことを目指す（貧困をなくす、飢餓をゼロに、伝染病の根絶、新生児および5歳未満児の予防可能な死亡の根絶など） ・さらに野心的：「誰も置き去りにしない」 ・大胆で変革的

際8つの目標のうち、7つは途上国のための目標でした。SDGs は普遍的な目標であり、先進国と途上国を含むすべての国のための目標です。

(4) MDGs はトップダウンのプロセスで準備作成されました。SDGs の場合は、政府や市民社会に加え、多様なアクターとの協議を重ね、世界各地のさまざまな声を聞き、より参加型なアプローチによって準備作成されました。

(5) MDGs は貧困状況にいる人々や他の開発状況の「削減」を目標としました。SDGs はこのような状況にいる人々を「なくす」という目標を掲げています。貧困の撲滅、飢餓に終止符を打つ、新生児および5歳未満の子供が予防可能な理由で死亡する数をゼロにする、などがその例です。

(6) この比較でお分かりになるように、SDGs は野心的な目標です。広範囲な分野で、さまざまな状況を終わりにし、そして「誰も置き去りにしない」ことを宣言しています。

(7) SDGs は大胆で変革的(transformative)です。SDGs に掲げられている目標を達成するためには、かなり踏み込んだ変革や改革が必要です。

3 2030 アジェンダ/SDGs の実施について

2030 アジェンダ/SDGs の実施は 2016 年 1 月に始まりました。2030 アジェンダでは、すべての国々が貧困撲滅、不平等の是正、平和で包括的で強靭な社会の構築、そして地球を守り、将来の世代の安寧を確保することが求められています。そしてすべて

の目標の進展と達成のためにはジェンダー平等の実現と女性・女児のエンパワーメントが非常に重要です。

　SDGsはグローバルなターゲットであり、すべての国の積極的な参加が求められています。SDGs実施の責任は各国にあります。各国政府はグローバル・レベルのSDGsターゲットを踏まえつつ、自分の国の状況を見据え、自国のターゲットを設定して達成することが必要です。

　2030アジェンダのフォローアップとレビューは、様々な国の異なる状況、能力や開発のレベルを考慮した上で、自主的に、それぞれの国が主導するものであるとされています。

　SDGs実施については次のようなレビューと報告があります（図2参照）。

(1) 1点目は、各国が作成する「自発的国家レビュー（Voluntary National Review: VNR）」です。その名のとおり、これは各国の主導と意思のもとに作成されます。それぞれの国が、自分の国のSDGs状況をまとめたVNR報告書を国連事務局に提出し、これが国連経済社会理事会（United Nations Economic and Social Council: ECOSOC）の主催による「持続可能な開発に関するハイレベル政治フォーラム（High-Level Political Forum on Sustainable Development: HLPF）」で話し合われます。報告書の形式については、国連が発行した共通のガイドラインはありますが、報告書の内容や表現は各国が決めます。これら国家レビューの提出と話し合いによって、各国の状況や、成功例、課題、教訓を含む経験がSDGs達成に向けて共有され、進捗を加速

図2：2030アジェンダ/SDGs の実施―報告とレビューについて

報告書・レビュー	作成者	頻度
1. 自発的国家レビュー（Voluntary National Review：VNR）	各国（自発的に）	VNR は毎年開催される HLPF に提出され、話し合われる
2. 持続可能な開発目標報告書（Sustainable Development Goals Report）	国連事務総長	毎年
3. 持続可能な開発グローバル報告書（Global Sustainable Development Report）	科学者・専門家で構成される独立したグループ（国連の作業チームのサポートを得て）	4年ごと

・これら3つの報告書はそれぞれが独立したものであるが、内容は補完的。
・それぞれが異なる視点からの報告をすることで ECOSOC 主催による「持続可能な開発に関するハイレベル政治フォーラム（High-Level Political Forum on Sustainable Development：HLPF）」でのレビューに貢献。

することがその目的とされています。

2016年には22カ国が国家レビューを HLPF に提出しました。2017年には、日本を含む44カ国が国家レビューを提出する予定です。

(2) 2点目は国連システムの協力の下、国連事務総長が毎年作成する「持続可能な開発目標報告書（Sustainable Development Goals Report）」です。この報告書ではグローバルな指標フレームワークおよび各国の統計・情報によって作成されたデータにもとづき、SDGs の進捗状況がまと

められており、深刻な諸問題・課題にも焦点があてられています。
(3) 3点目が「持続可能な開発グローバル報告書 (Global Sustainable Development Report)」です。この報告書は、持続可能な開発を促進するための科学と政策の協調を強化することを目的として科学者・専門家で構成される、独立したグループによって作成されます。国連の作業チームのサポートを得て、4年に一度、包括的で綿密な報告書を作成します。

これら3つの報告書はそれぞれが独立したものですが、内容は補完的です。HLPFでのレビューでは、それぞれが異なる視点からの報告をすることで貢献します。HLPFはSDGs実施において政治的なリーダーシップを持ち、指針、提言を提供します。2030アジェンダ/SDGsのフォローアップとレビューをグローバル・レベルで監督することにおいて、HLPFは中心的な役割を担っています。HLPFには国連の全加盟国が参加します。加えて、市民社会団体の代表なども参加することができます。

ただしHLPFには決定する権限はありませんし、加盟国間や国連機関の間を調整する役割も与えられていません。HLPFは持続可能な開発に関する経験やベストプラクティスを多くのステークホルダーの間で共有することを促進します。また、持続可能な開発政策に関する国連システム全体の一貫性を推進します。

4 課題

　それでは 2030 アジェンダ/SDGs の実施においては、どのような課題があるのでしょうか。時間の関係で全部の課題についてここでお話することはできませんので、いくつかの点に絞ってお話いたします。

(1) SDGs についての意識を高め、理解を深める：SDGs に含まれる分野は幅広く、多くの課題が複雑に絡み合っているので、その内容と意義を世界の人々に効果的に伝えるための広報戦略が必要です。このグローバル・アジェンダが、世界の一人ひとりにとって意義のあるものであり、自分自身に関係のあるものだと捉えられることが重要です。そのためには、例えば学校の教科書に SDGs についての記述を含めます。また、従来のメディアでの伝達・広報に加えてソーシャル・メディアを積極的に使うことも必要だと思います。

(2) 多様なアクターの参画：SDGs 実施のためには、多様なアクターの参画が不可欠です。そのためには、トップレベルの強い政治的コミットメントとリーダーシップが必要となります。多様なアクターの中には政府組織、国会や地方議会の議員、市民社会団体、民間企業、学術・研究機関、メディア、自治体やコミュニティーなども含まれます。これ

らの異なるアクターが継続的に協力してSDGs実施に参画するためには組織的な枠組みの設置が必要です。例えば、政府首脳をリーダーとして関係省庁と異なるアクターの代表で構成されるSDGs実施促進委員会を設置する。多くの国では、その能力と経験の限界とも関連して、地方政府・自治体の積極的な参画が課題のひとつです。SDGs実施のためには「whole-of-government approach（政府全体としてのアプローチ）」が必要であることは当然ですが、それ以上に「whole-of-society approach（社会全体としてのアプローチ）」で社会を総動員するアプローチが必要とされます。

2030アジェンダとSDGs実施において、私が大きく期待するのが若者の役割です。今日の世界では、3人にひとりは30歳以下です。これら若者の可能性を引き出し、活動・活躍の機会を広げることが重要です。若者には、力強いパワーがあります。そして変革をもたらすことができます。彼らはパワフルなチェンジ・エージェントなのです。この会場にも、多くの若者がいらっしゃるので、皆様に期待しています。

(3) SDGsの主流化：SDGs実施のためには、これら目標が国家を含むさまざまな組織、そして人々の活動の中に盛り込まれていなければなりません。言いかえれば、SDGsの主流化が必要だということです。つまり、特別なこととしてSDGs活動を行うのではなく、通常の活動・仕事として

SDGs を実施するのです。このためには、国全体としての開発計画の他に各省庁ごとの計画や、地方自治体の行動計画などに、SDGs が戦略・政策の基本的な要素として含まれていることが大事です。SDGs は、政府組織や公的機関だけでなく、民間企業や市民社会団体などを含むすべての組織の政策や活動計画にも反映されることが重要となります。そして、これらに基づき SDGs 実施のための予算を確保するのです。

(4) 目標の優先度に関する課題：国連総会で採択された文書には、SDGs は包括的で不可分だと明確に述べられています。SDGs に含まれる 17 の目標を、どのように包括的に、同時に優先課題やニーズを踏まえて実施するのかは十分に考慮されなければなりません。これに関しては、各国でさまざまな分析方法や取り組みが進められています。たとえば、ギャップ分析にもとづいて目標の優先度を決めて国家計画に反映する取組があります。多くの目標とターゲットは関連していますので、シナジー効果を最大限に高める形で実施を進めることが有効でしょう。

(5) データの入手：SDGs に含まれる 17 の目標と 169 のターゲットの進捗を計るためのグローバル・レベルの指標は 200 以上になるのではないかと言われていますが、これについてはまだ議論が続いており、最終的な指標は合意されていません。[1]ちなみに MDGs には 8 つの目標、18 のターゲット、

60 の指標が含まれていました。どちらにしても、SDGs のために必要となる統計を確保することは各国の総計機関と国際的な統計関係者たちにとっては、いままでにない挑戦となることでしょう。

　現状の把握と SDGs の進捗状況を計るためには常に最新の信頼できる統計が必要となります。そして、その統計は男女、年齢、所得、人種、移民ステータス、障害の有無、地理、等の特徴によって分類されていることが必要です。これら異なるグループ間の格差や不平等を計り、分析しなくてはならないからです。ところが、そのように分類整理された統計は多くの国では存在しません。統計の収集や分析を行うための人材も十分いなければ、その能力もない国が多いのが現状です。最も脆弱な立場に置かれている農村地域の女性、先住民族、スラムの住民、紛争下の人々などは、今までも統計から漏れてしまっているのです。各国のデータの収集、分類、分析、共有のための能力強化は必須です。

(6) ターゲットと指標：統計と密接に関係しているのが指標です。2015 年に 2030 アジェンダが採択された時には、SDGs の目標とターゲットは設定されましたが、これらの進捗を計るための指標は合意されませんでした。各目標・ターゲットのグローバル・レベルの指標フレームワークは、国連統計委員会（United Nations Statistical Commission）により設置された「持続可能な開発目標（SDGs）指標に

関する機関間専門家グループ（Inter-Agency and Expert Group on Sustainable Development Goal Indicators）」が策定することになっています。国連統計委員会で合意されたグローバル・レベルの指標フレームワークは、その上部組織である ECOSOC と国連総会に提出され、承認されるという仕組みです。

2016年3月の統計委員会会合は、この専門家グループが提示した SDGs 指標を「実質的な始まり（"practical starting point"）」としました。1年後の今月、2017年3月に専門家グループは SDGs 指標の修正版を統計委員会に提出しました。この状況でお分かりになるように、この課題に関しては審議が続いています。[2]

グローバルな SDGs 指標を設定するのは複雑な課題です。国連統計部は国際的に比較可能である指標の重要性を強調しています。「国内および国家間の不平等を是正する」（目標10）や「持続可能な開発に向けて平和で包摂的な社会を推進する」（目標16）のように適切な指標について国際社会の合意を得るのが困難な目標もあります。

(7)「誰も置き去りにしない」状態をどのように達成するのかは大きな課題です。最も脆弱な立場に置かれた人々や危機的な状況にいる人たちにはどのようにアクセスしたらよいのでしょうか。特に考慮されなければならないのは次の人々です。
　― 難民、国内避難民、移民

― 女性
― 子供
― 障害者
― 先住民族
― 高齢者（多くの途上国の最も貧しい人たちの中には多くの高齢者が含まれています）
― 都市部の住民（世界で最も貧しく疎外されている人の多くが都市部に住んでいます）

　これらの人々に関しては、不平等、排除、差別の要因をつきとめて、対処しなければなりません。貧困層から一度は抜け出した人たちが、なぜ再び貧困層に戻ってしまうのか。なぜ貧困状態から抜け出すことができないのか。この人たちの脆弱性はどのようにしたら軽減でき、どのようにしたら強靭性を高めることができるのでしょうか。どのような戦略や政策が彼らの状況を改善することができるのでしょうか。「誰も置き去りにしない」ためには多くの課題に取り組まなければなりません。

(8) SDGs 実施のための資金：2030 アジェンダが成功するかどうかは、どれだけの資金を調達できるかによります。世界のリーダーたちは持続可能な開発のための資金のグローバルな枠組である「アディス・アベバ行動目標」に合意しました。そこでは SDGs 実施のための公的・民間資金、国内・国際資金の調達のあらゆる可能性を探ることが公約されました。国内での資金調達が必要なことは言うまでもありま

せん。各国はそのために、徴税体制の改善、税金逃れや違法な資金流通の取り締まりと防止に力を入れ、資金源を広げることに合意しています。政府開発援助（Official Development Assistance: ODA）は当然必要であり、それを最も必要としている国に向けることが求められています。民間投資はより拡大され、持続可能な開発のために向けられるべきです。さまざまな形の革新的な資金調達の可能性も追求されるべきでしょう。

2030アジェンダを実施するために必要な資源・財源は世界に存在するのです。この存在する資源をどのようにして持続可能な開発活動に向けるのかが大事な点であり、それによってSDGsが達成されるかどうかが決まるのです。

5 SDGs達成のための国連の役割

それでは、SDGs達成においての国連の役割について考えてみましょう。

(1) 1点目として、国連システムの諸機関は、その特徴と強みであるグローバルな組織とネットワークを生かして、それぞれの専門分野で2030アジェンダの実施を積極的に支援することができます。これには、グローバル、地域、国レベルの活動が含まれます。

(2) 2点目は、国連事務総長の呼びかけ「（国連）憲章のすべ

ての部分（work across the United Nations Charter）」で活動を進めることです。2030アジェンダには国連の基本的な使命と価値観が反映されています。SDGsに含まれる課題の範囲は広く、諸々の要素が複雑に絡み合い、密接に繋がっているので、包括的なアプローチが必要とされます。国連システムの諸機関の専門性、資源、経験、知見、グローバル・ネットワークを総動員して開発、人道、平和と安定、人権のすべての分野の活動を包括的に進めることが重要です。

(3) 3点目は、国際社会で合意された規範や基準を国連が擁護することです。脆弱な立場に置かれている人々、あるいは疎外されている人々のニーズに取り組むためにはこの点は重要です。

(4) 4点目は、グローバルな合意をまとめ、さまざまなアクターの協働を推進することです。2030アジェンダには気候変動、紛争、難民問題、感染症など多くの地球規模課題が含まれており、その解決のためには、すべての国のコミットメントと協調した活動・行動が必要です。国連はその招集機能を用いて、多様なアクターがSDGsについて協議できる場を提供することが求められています。これらの協議を通じ、国際社会の合意やグローバル・パートナーシップの形成を促進することが国連の重要な役割です。

(5) 5点目は2030アジェンダとSDGsの進捗のモニタリングです。SDGsに関する報告書については既にお話いたしましたが、HLPFはSDGs実施に関し、政治的リーダーシップを発揮し、指針・提言を話し合うためのグローバルな場を提供しています。国連諸機関はグローバル・レベルの報告書の作成に参加、あるいはそれぞれの組織の専門分野での報告を作成しています。同時に、国連諸機関は各途上国に対する支援も実施しています。

(6) 国レベルでは、各国自身のオーナーシップ、主体性が最も重要です。国連の支援は、それぞれの国のニーズ、優先課題、能力などにもとづいて提供されます。多くの途上国はSDGs実施における国連の支援をすでに要請しています。

30以上の開発分野の国連機関で構成される国連開発グループ（United Nations Development Group）は、国連加盟国の要請に応え、「MAPS approach」を提供しました。MAPSはMainstreaming（主流化）、Acceleration（加速化）、そしてPolicy Support（政策支援）の意味です。

「Mainstreaming（主流化）」は、SDGsを国家と地方の戦略・計画・予算に組み込むことです。「Acceleration（加速化）」は、主流化された活動分野のうち、優先度が高いものに資源や予算を集中させて、その進捗を加速させることです。加速化のためには、進捗を妨げている問題は何なのか、どのようにしたらシ

ナジー効果が高まるのか、必要な資金をどのように調達するのか、どのようなパートナーとの連携が効果的なのか、などを明確にしなければなりません。このような分析にもとづき、SDGs実施を効果的に行うことが大事です。国連の諸機関は、ニーズに応じて「主流化」と「加速化」のための支援を行います。3番目の「Policy Support（政策支援）」は、国連機関の専門性と長年の経験・知見に基づいて提供されます。

　以上、6つの国連の役割を挙げました。これでおわかりになりますように、国連がSDGs実施に関して果たす役割は、サポーター（supporter）、ファシリテーター（facilitator）、プロモーター（promoter）、そしてアクセルレーター（accelerator）です。同時に、国連は、アイディアのイニシエーター（initiator）、新しいアイディアやアプローチのクリエーター（creator）、注意の呼びかけ人（attention-caller）、そして啓発者（advocator）でもあります。

6　SDGs達成のための日本の役割

　SDGs達成に関しての日本の役割は国内的な面と国際社会での役割と2つあります。

　日本国内では2016年、内閣に「持続可能な開発目標（SDGs）推進本部」を設置しました。[3]SDGsの実施を推進・モニタリングするため、内閣総理大臣を本部長とし、全閣僚を構成員としています。同じ年には「持続可能な開発目標（SDGs）実施指針」

を定めました。[4] この実施指針は、日本が 2030 アジェンダの実施にかかる重要な挑戦に取り組むための国家戦略です。

この「持続可能な開発目標（SDGs）実施指針」の概要のスライド（図3）をご覧いただいていますが、ビジョン、実施原則、8つの優先課題とフォローアップについてまとめられています。

8つの優先課題は、「SDGsのゴールとターゲットのうち、日本として特に注力すべきものを示すべく（中略）、すべての優先課題について国内実施と国際協力の両面が含まれる」と述べられています。

時間の関係で、各項目についての説明はここではできませんが、最初の優先課題「あらゆる人々の活躍の推進」に触れさせていただきます。この課題の中には「女性活躍の推進」が含ま

図3

持続可能な開発目標（SDGs）実施指針の概要
- ビジョン：「持続機能で強靭、そして誰も置き去りにしない、経済、社会、環境の統合的向上が実現された未来への先駆者を目指す。」
- 実施原則：①普遍性、②包摂性、③参画型、④統合性、⑤透明性と説明責任
- フォローアップ：2019年までを目処に最初のフォローアップを実施。

【8つの優先課題と具体的施策】

①あらゆる人々の活躍の推進	②健康・長寿の達成
■一億総活躍社会の実現■女性活躍の推進■子供の貧困対策■障害者の自立と社会参加支援■教育の充実	■薬剤耐性対策■途上国の感染症対策や保健システム強化、公衆衛生危機への対応■アジアの高齢化への対応

③成長市場の創出、地域活性化、科学技術イノベーション	④持続可能で強靭な国土と質の高いインフラの整備
■有望市場の創出■農山漁村の振興■生産性向上■科学技術イノベーション■持続可能な都市	■国土強靭化の推進・防災■水資源開発・水循環の取組■質の高いインフラ投資の推進

⑤省・再生可能エネルギー、気候変動対策、循環型社会	⑥生物多様性、森林、海洋等の環境の保全
■省・再生可能エネルギーの導入・国際展開の推進■気候変動対策■循環型社会の構築	■環境汚染への対応■生物多様性の保全■持続可能な森林・海洋・陸上資源

⑦平和と安全・安心社会の実現	⑧SDGs実施推進の体制と手段
■組織犯罪・人身取引・児童虐待等の対策推進■平和構築・復興支援■法の支配の促進	■マルチステークホルダーパートナーシップ■国際協力におけるSDGsの主流化■途上国のSDGs実施体制支援

出典：持続可能な開発目標（SDGs）推進本部

れていますが、特に踏み込んだ行動が必要なことは確実です。皆様ご存知かと思いますが、世界経済フォーラムが毎年発表している「グローバル・ジェンダー・ギャップ指数（Global Gender Gap Index）によると、2016年の日本のランキングは144カ国中111位でした。

　国際社会での日本の役割について実施指針は、「我が国は、このような持続可能な経済・社会づくりに向けた先駆者、いわば課題解決先進国として（後略）」、「今後のSDGs実施段階においても、世界のロールモデルとなることを目指し、国内実施、国際協力の両面において、世界を、誰も置き去りにされることのない持続可能なものに変革するための取組を進めていくことを目指す」と述べています。

　日本は国際協力において、保健衛生、防災、ジェンダー平等などに力を入れてきました。また日本のODAは長年の間、多くの途上国を支援してきました。ODAは、国際協力の重要な柱として今後も続けられるべきでありますし、特に最貧国、紛争や自然災害で危機的状況にある国々に向けられることが期待されます。日本は南々協力や三角協力を進めてきていますが、これら支援の今後の拡充も期待しています。

　日本は2030アジェンダの準備作成に積極的に関わってきました。実施段階においても日本が活発に行動し、国際社会でリーダーシップを発揮することを期待しています。日本は2017年に自発的レビュー報告書を国連に提出する予定ですが、その内容が楽しみです。

　日本の民間企業が今よりもさらに活発にSDGsに関わること

を期待します。これは国内と海外での両方です。SDGs は民間企業にとって、技術、経営ノウハウなどのビジネス経験に基づいた持続可能な解決策を提供できる機会なのです。そして、それを企業の中核的なビジネスとして実施することも期待できます。

若者のパワーについて先ほど触れましたが、日本の若者にはもっと積極的に SDGs 実施に関わって欲しいですし、関わるべきです。社会における女性の役割はもっと高まることが必要であり、SDGs に関わる女性も増えることを期待します。他のアクター全員が SDGs 達成に向けて積極的に行動することが必要です。この 3 つのグループを挙げたのは、SDGs 実施への参画にとても大きな可能性が秘められているからです。

今年 2017 年は、SDGs 実施の 2 年目です。その達成に向けて私たちは行動を加速しなければなりません。世界で SDGs 達成のために活動しているアクターは膨大な数にのぼります。アクター同士が協力し、お互いから学び合い、そうすることによって活動を加速化させ、包括的な取り組みを推進して SDGs 達成に向けて進むことが重要です。

今、大きな機会が私たちの目の前にあります。私たちはそのチャンスを捉えなくてはなりません。皆様が SDGs 実施に参加して行動をとることによって、世界の人々の状況をよりよいものにできるのです。日本、そして海外で、皆様が積極的に行動すること、そして様々なパートナーと協力して活動することが、2030 アジェンダと SDGs 達成へと進むことなのです。

さあ皆さま、SDGs 達成に向けて行動をとりましょう！今こ

そが行動に参加する時です！

注

1 2017年7月に国連総会で採択されたSDGsグローバル指標フレームワークには230の指標が含まれています。
2 2017年3月に「SDGs指標に関する機関間専門家グループ」が策定した指標フレームワークは国連統計委員会に同月採択され、さらに上部組織である国連経済社会理事会（ECOSOC）で6月に、そして国連総会で7月に採択されました。これによってSDGsの17の目標について230の指標が設定されました。
3 持続可能な開発目標（SDGs）推進本部： http://www.kantei.go.jp/jp/singi/sdgs/
4 持続可能な開発目標（SDGs）実施指針：http://www.kantei.go.jp/jp/singi/sdgs/dai2/siryou1.pdf

1 The Roles of UN and Japan in Achieving the Sustainable Development Goals (SDGs)

Akiko Yuge

First, I would like to thank the organizers of this symposium, namely the United Nations University Institute for the Advanced Study of Sustainability and the Japan Foundation for the United Nations University, for inviting me to give a keynote lecture at this important symposium.

It's wonderful that UNU Alumni Association is having its tenth anniversary. I would like to express my congratulations to the members of the Alumni Association and all those concerned with the organization of this symposium to commemorate this special occasion.

I am also conscious that today marks the sixth anniversary of the terrible 3.11 triple disaster of earthquake, tsunami, and nuclear meltdown. We should recall what happened on that day, and think deeply about the issues involved – disaster, recovery, resilience, and other aspects.

The outline of my lecture is as follows. First, I will briefly review the major global agreements reached in the last two years, so as to place today's lecture in context. I will then describe the Sustainable Development Goals (SDGs), comparing them to the Millennium Development Goals (MDGs). This will be followed by the implementation of the 2030 Agenda for Sustainable Development, and related issues and challenges. I will then talk about the role of the United Nations (UN) and Japan in achieving the SDGs.

1 Major Global Agreements in 2015 and 2016

So, what were the major global agreements reached during the last two years? Here is a list of the global agreements and conferences.

The Sendai Framework for Disaster Risk Reduction was agreed here in Japan in March 2015. This 15-year framework, covering the same period as the 2030 Agenda, aims to reduce disaster risk, and losses in lives, livelihoods, and health of people.

In July 2015, the Addis Ababa Action Agenda was agreed at the Third International Conference on Financing for Development. The meeting agreed to draw on all possible sources of funds for development – including public and private funding, domestic and international finance, as well as other innovative financing arrangements.

At a summit meeting at United Nations Headquarters in September 2015, world leaders agreed on the 2030 Agenda for Sustainable Development and its 17 SDGs.

In December 2015, the Paris Agreement on climate change was reached. It was a landmark decision to combat climate change and adapt to its effects.

In May 2016, the World Humanitarian Summit was held in Istanbul, and this led to multiple commitments and initiatives to strengthen humanitarian and development actions to reduce suffering of people in crisis, disasters, and forced displacement.

The subsequent United Nations Summit on Addressing Large Movements of Refugees and Migrants in September 2016 made commitments to protect the human rights of all refugees and migrants.

Then, in October 2016, the United Nations Conference on Housing and Sustainable Urban Development (known as HABITAT III) was held in Ecuador. The New Urban Agenda was adopted setting a course towards sustainable urban development by rethinking how cities are planned, managed, and inhabited.

Noteworthy is the fact that all these global conferences and agreements were sponsored by the United Nations to call attention, discuss, and forge global agreements on issues of vital importance. The outcomes of these meetings form an integral part of the 2030 Agenda.

2 Comparison between the MDGs and the SDGs

As you are aware, the MDGs preceded the SDGs. The SDGs will build upon the achievements of the MDGs and seek to address their unfinished business. The SDGs are comprehensive, far-reaching, and go far beyond the MDGs.

Let us look at the key differences between the MDGs and the SDGs. Please see Figure 1 for this comparison.

(1) The MDGs had 8 goals with 18 targets; the SDGs have 17 goals with 169 targets.

(2) The MDGs focused on social sector issues, whereas the SDGs are much broader and comprehensive, covering economic, social, and environmental dimensions of sustainable development.

Figure1 Comparison between MDGs and SDGs

MDGs	SDGs
· 8 goals with 18 targets · Focus on social sector · Focus on developing countries (7 out of 8 goals) · Top-down preparation process · Aim to "reduce"	· 17 goals with 169 targets · Comprehensive coverage for sustainable development: economic, social, environmental · Universal goals for all countries · Consultative and participatory preparation process · Aim to "end" (poverty, hunger, epidemics, preventable deaths of newborns and children under 5 years of age) · More ambitious: "leave no one behind" · Bold and transformative

(3) The MDGs' goals and targets focused on developing countries (seven out of eight goals were for developing countries). The SDGs are universal goals for all nations, developing and developed.

(4) The MDGs were prepared through a top-down process. For the SDGs, the preparation process was more consultative and participatory, and engaged governments, civil society, and other stakeholders around the world.

(5) The MDGs aimed to "reduce" extreme poverty and other human conditions. The SDGs aim to "end" poverty, hunger, preventable deaths of newborns and children under five years of age, and more.

(6) The SDGs are ambitious, with its comprehensive coverage, goals to end poverty and other conditions that I just mentioned, and its aim to "leave no one behind".

(7) The SDGs are transformative – to take bold and transformative steps and reforms to achieve these ambitions.

3 Implementation of the 2030 Agenda and the SDGs

Implementation of the 2030 Agenda started in January 2016. So, it's been just over a year since its start. The achievement of the 2030 Agenda requires all countries to: eradicate poverty; combat inequalities; build peaceful, inclusive and resilient societies; protect our planet; and secure the well-being of future generations. Gender equality and empowerment of all women and girls are central to the achievement of all goals.

Targets are defined as aspirational and global, with each government setting its own national targets. Primary responsibility for implementing the SDGs rests with nation states. Governments should be guided by the global level of ambition, while taking into account their own national circumstances.

The 2030 Agenda states that follow-up and review processes will be "voluntary and country-led", and "will take into account different national realities, capacities and levels of development". Progress in the SDGs implementation is being reported and

Figure 2 Implementation of 2030 Agenda: Reporting and Reviews

Reports/reviews	Prepared by	Frequency
1. Voluntary National Review (VNR)	Member States (on a voluntary basis)	VNR presented to HLPF are reviewed annually at its meeting
2. Sustainable Development Goals Report	United Nations Secretary-General	Annual
3. Global Sustainable Development Report	Independent group of scientists and experts (supported by UN task team)	Every 4 years

・Above reports are distinct but complementary
・They all contribute to discussion at High-Level Political Forum on Sustainable Development (HLPF) that meets under the auspices of UN ECOSOC

reviewed in several ways. I will mention three (see Figure 2).

(1) The first is the Voluntary National Review (VNR): As the name indicates, these reviews are voluntary, and it is country-led. Member States volunteer to prepare their own national reviews and present them for discussion to the High-Level Political Forum on Sustainable Development (HLPF) that meets under the auspices of United Nations Economic and Social Council (ECOSOC).

While the UN provided a set of common reporting guidelines, each country decides on the scope of the national review and its format. These national reviews aim to facilitate the sharing of experiences among countries, including successes, challenges and lessons learned, with a view to accelerating the implementation of the 2030 Agenda.

In 2016, 22 countries presented Voluntary National Review to this forum. In 2017, 44 countries, including Japan, volunteered to present their national reviews.

(2) The second is the Sustainable Development Goals Report prepared by the UN Secretary-General on an annual basis. It aims to present a global overview of the SDGs implementation to highlight the most significant gaps and challenges.

(3) The third is the Global Sustainable Development Report, prepared by an independent group of scientists and experts, supported by a UN task team. This report focuses on the science-policy interface of sustainable development. A

comprehensive, in-depth report will be produced every four years.

The three reports that I mentioned are distinct but complementary in nature. They contribute different perspectives to the discussions at the High-Level Political Forum on Sustainable Development (HLPF). This forum provides political leadership, guidance, and recommendations for the implementation of sustainable development commitments. As such, it has a central role in overseeing follow-up and review of the 2030 Agenda and the SDGs at the global level. It is a forum in which all UN Member States can take part. In addition, representatives of civil society organizations have options to participate.

It should be noted that the forum neither has any concrete decision-making powers nor does it perform a direct coordinating role vis-à-vis governments and UN organizations. It primarily serves the purpose of facilitating the sharing of experiences and best practices and promote system-wide coherence and coordination of sustainable development policies.

4 Issues and challenges

So, what are the issues and challenges in the implementation of the 2030 Agenda and the SDGs? Let me raise several of them. Please bear in mind that this list is not exhaustive due to time limitation.

(1) Awareness raising: Considering the very broad scope and complexity of the SDGs, there must be effective communication strategies, awareness-raising, advocacy, and outreach activities to make the SDGs more understandable to a broad audience. The global agenda needs to be translated into language that is meaningful to different target groups. For instance, school curricula should now include the SDGs. In addition to the traditional media, active use of social media should also be promoted.

(2) Involvement of multiple stakeholders: For a broad range of stakeholders to be involved in the implementation of the SDGs, strong political commitment and leadership from the highest level is necessary. The stakeholders include government ministries and agencies, parliamentarians, civil society organizations, the private sector, academic institutions, the media, communities, and others. Moreover, a clear institutional structure for multi-stakeholder engagement is needed. For instance, establishing an inter-ministerial committee or multi-stakeholder commission on the SDGs, led by the head of government, would be an example. For many countries, local government engagement remains a challenge, partly because of their limited capacity. A "whole-of-government approach" is needed; and moreover, an inclusive "whole-of-society approach

to the SDGs" is needed.

The youth must be an active player in the implementation of the 2030 Agenda. Actually, one in every three people today is under the age of 30. Their potential should be tapped and opportunities created. Young people are powerful drivers of change. And I can see that we have many young people in this hall today.

(3) Mainstreaming the SDGs: The SDGs have to be translated and embedded into the work of nations, organizations, and people. The SDGs must be mainstreamed – meaning incorporated into the strategies and policies of national development plans, sectoral plans, and local government plans. This should be the case, not only for government agencies, but also for civil society organizations, private sector firms, and other bodies. Budget should be allocated accordingly for the implementation of the SDGs.

(4) Determining priorities and balancing goals: The SDGs and their targets are integrated and indivisible. How best to implement the SDGs in an integrated manner while balancing different demands and priorities has to be carefully considered. Countries have been using different analytical tools, such as gap analysis, to define priorities and goals, and reflect them into national plans. As the goals and targets are interlinked, synergies among them should be maximized.

(5) Data availability: Data requirements for the SDGs with

their 17 goals, 169 targets, and more than 200 global indicators that are yet to be finalized, present an unprecedented challenge for both national statistical systems and the international statistical community.[i] You may recall that the MDGs had 8 goals, 18 targets, and 60 indicators. Accessible, timely, high-quality, and reliable data are necessary to assess the situation and measure progress.

What is needed is disaggregated data by sex, age, income, ethnicity, migration status, disability, geographic location, and other characteristics to be able to analyze gaps and inequalities. Many countries neither have the necessary data nor the human or financial resources or capacity to collect such data and analyze them. Many of the most vulnerable population, including rural women, indigenous peoples, people living in slums, and people affected by conflict, are consistently left out of data sets. Building national capacities for data collection, disaggregation, dissemination, and analysis is essential.

(6) Targets and indicators: issue of measurement: This is a related point – the indicators. While the goals and their targets were decided when the 2030 Agenda was adopted in 2015, the specific indicators to measure progress of each goal and target have not been agreed at that time. The task to develop global indicators was given to the Inter-Agency and Expert Group on Sustainable Development Goal Indicators. This Group was established by the United Nations Statistical Commission in

2015. Once agreed at the United Nations Statistical Commission, the global indicator framework has to be submitted for adoption at ECOSOC and the UN General Assembly.

In March 2016, the Statistical Commission accepted the proposed SDGs indicator framework submitted by this Group as "practical starting point". A revised indicator framework was submitted to the UN Statistical Commission for adoption this month. As such, this is work in progress.[ii]

This work is actually quite complex as the Statistical Commission emphasizes the importance of "guaranteeing international comparability" of indicators. Then, there are challenges to agree on appropriate indicators for such goals as "Reduce inequality within and among countries" (Goal 10), and "Promote peaceful and inclusive societies for sustainable development..." (Goal 16).

(7) Achieving "no one will be left behind": A major challenge is how to reach the most vulnerable and most further behind, including those in fragile states and crisis situations. The needs of the following groups of people must be given special attention:

- refugees, internally displaced people, and migrants;
- women;
- children;
- persons with disabilities;

- indigenous people;
- older people, who are among the poorest in many developing countries;
- urban population, since many of the world's poorest and most marginalized people now live in cities.

At the same, the drivers of inequality, exclusion, and discrimination must be identified and tackled for these groups. Why do people fall back into poverty and cannot get out? How can their vulnerability be reduced and their resilience strengthened? What type of strategies and policies would be effective to improve their situation? These are some of the questions that need to be looked into.

(8) Financing for the SDGs implementation: Financing is the linchpin for the success of the 2030 Agenda. World leaders agreed to the Addis Ababa Action Agenda that provided a global framework for financing sustainable development. It calls for public and private, domestic and international funds to implement the SDGs. Domestic resource mobilization is central to the financing agenda. Countries agreed to widen the revenue base, improve tax collection, and combat tax evasion and illicit financial flows. Official Development Assistance (ODA) flows must continue to help the countries most in need. More private investment, and at a much larger scale, should be aligned with sustainable development. Innovative financing should also be explored.

The resources exist in the world to finance the 2030 Agenda. How those resources will be directed to support sustainable development will be the key to achieving the SDGs.

5　Role of the UN in Achieving the SDGs

Now, we will move to the role of the UN in achieving the SDGs.

(1) Engage and support at global, regional, and national levels: First, with its global presence, the United Nations system and its agencies have the unique strength to actively engage and support the implementation of the 2030 Agenda at global, regional, and national levels.

(2) "Work across the (UN) Charter": Secondly, the UN Secretary-General has called for all of us to "work across the Charter"– the UN Charter. Indeed, the 2030 Agenda speaks to the core mandate and values of the UN System. The broad coverage, complexity, and interlinkages of the SDGs require an integrated approach. The expertise, resources, experience, and the global network of the UN System agencies across development and humanitarian, peace and security, and human rights should be brought together to advance the SDGs.

(3) Uphold internationally agreed norms and standards: Thirdly, the UN will uphold internationally agreed norms and

standards to serve the needs of the people, especially those that are most vulnerable and marginalized.

(4) Broker global agreements for collective action: Many of the global challenges contained in the 2030 Agenda, such as climate change, violent conflicts and refugee crisis, and contagious diseases require shared commitment and collective action across nations. Here, the UN should use its convening power to organize conferences to bring diverse partners to discuss various aspects of the SDGs as the situation requires. Through these discussions, the UN has a key role in brokering global partnerships and agreements to accelerate collective action.

(5) Monitor progress of the 2030 Agenda and the SDGs: The UN has an important role in monitoring implementation of the 2030 Agenda. Earlier, I talked about the various SDGs reports, and these are prepared by, supported by, and/or presented to and discussed at UN bodies.

(6) At the country level, national ownership is paramount. Support by UN agencies is based on the needs, priorities, and capacity of each country. Many developing countries have already requested support of the UN Country Team for the SDGs implementation.

In response to requests by Member States, the United Nations Development Group (UNDG), comprising over thirty UN agencies that work in the development field, came up with the "MAPS approach". MAPS stands for Mainstreaming, Acceleration, and Policy Support. Let me explain.

"Mainstreaming" is to incorporate the SDGs into the country's national and local strategies, plans, and budgets. "Acceleration" means targeting resources to priority areas identified in the "mainstreaming" process. "Acceleration" also includes identification of constraints, synergies and trade-offs across sectors, means of financing, and partnerships potentials. Based on such an analysis, how best to implement the SDGs will be decided. In "mainstreaming" and "acceleration", UN agencies can provide support where it is needed. The third pillar, "policy support" is available by UN agencies using their expertise and extensive experience.

In all these six points that I listed, the UN has an important role of being a supporter, facilitator, promoter, and accelerator in the implementation of the SDGs. At the same time, the UN is also an initiator and creator of new ideas and approaches, attention-caller, and advocate on the SDGs.

6 Role of Japan in Achieving the SDGs

Now, moving to Japan's role in achieving the SDGs – this is both at the domestic level and international level.
(1) Domestically: At the domestic level, in 2016, Japan established the SDGs Promotion Headquarters within the

Cabinet. It is headed by the Prime Minister and composed of all ministers to lead implementation and monitor progress of the 2030 Agenda. Later in the same year, the SDGs Implementation Guiding Principles were issued.[iii] This represents Japan's national strategy to address major challenges for the implementation of the 2030 Agenda.

This slide (Figure 3) is the outline of Japan's SDGs Implementation Guiding Principles.

The vision, implementation principles, eight priority areas,

Figure3

The Outline of the SDGs Implementation Guiding Principles (provisional translation)

- Vision:Set out a vision for Japan to be the champion of sustainable and resilient society in which "no one is left behind." Japan intends to be a leader in creating a better future, in which the three dimensions of sustainable development, namely, economic, social, and environmental are improved in an integrated manner.
- Implementation Principles: (1) Universality, (2) Inclusiveness, (3) Participatory, (4) Integration, (5) Transparency and Accountability
- Follow-up cycle: Expected to conduct a first follow-up by 2019

【Eight Priority Areas and Policies】

(1) Empowerment of All People	(2) Achievement of Good Health and Longevity
■Realization of Dynamic Engagement of All Citizens ■Promotion of Women's Role in Society ■Measures against Child Poverty ■Assistance to People with disabilities for Social Participation and Self-reliance ■Promotion of Quality Education	■Measures against Antimicrobial Resistance ■Enhancing Developing Countries' Health Sector and Improving Their Public Health and Measures against Infectious Diseases ■Tackling the Issues Associated with Aging Populations in Asia
(3) Creating Growth Market, Revitalization of Rural Areas, and Promoting Technological Innovation	(4) Sustainable and Resilient Land Use, Promoting Quality Infrastructure
■Creating Markets with Potentials ■Revitalizing Villages around Seas, Mountains, and Farmlands ■Improving Productivity ■Science and Technology Innovation ■Sustainable City	■Creating Resilient Land and Promoting Disaster Risk Reduction ■Water Resource Development and Measures on Water Circulation ■Promoting Quality Infrastructure Investment
(5) Energy Conservation, Renewable Energy, Climate Change Measures, and Sound Material-Cycle Society	(6) Conservation of Environment, including Biodiversity, Forests and Oceans
■Introduction and Promotion of Renewable Energy ■Measures against Climate Change ■Establishing Reeveling-based Society	■Measures against Environmental Pollution ■Biodiversity Conservation ■Sustainable Use of Forest, the Oceans, and Land Resource
(7) Achieving Peaceful, Safe and Secure Societies	(8) Strengthening the Means and Frameworks of the Implementation of the SDGs
■Tackling Organized Crime, Human Trafficking, and Child Abuse ■Peacebuilding and Assistance for Reconstruction ■Promotion of the Rule of Law	■Multi-Stakeholder Partnership ■Mainstreaming SDGs inV International Cooperation ■Assisting Developing Countries to implement SDGs

(Source: SDGs Promotion Headquarters, Government of Japan)

and follow-up cycle are summarized there. The Guiding Principles state that the eight priority areas are those among the SDGs and targets that Japan should focus. It further states that these priority areas include both domestic measures and those to be implemented through international cooperation.

While there is no time to go through all of these items now, I would like to pick up one item. The first priority area "Empowerment of all people" includes "Promotion of women's role in society". This area definitely requires priority action. As you may be aware, according to the Global Gender Gap Index published by the World Economic Forum last year, Japan ranks 111 out of 144 countries.

(2) Internationally: The SDGs Implementation Guiding Principles state that "Japan should contribute to the world as leading solution provider by sharing its successes, experience, and lessons learned in building a sustainable economy and society at home." It further states "Japan aims to become a role model for the world in the implementation of measures to achieve the SDGs and will make efforts, both in Japan and in cooperation with other countries, to achieve sustainable societies worldwide where no one will be left behind."

Japan has placed such issues as health, disaster risk reduction, and gender equality at the core of its international cooperation. Japan has had a long history as ODA provider to a large number of developing countries. Japan's ODA must remain a very important pillar of international cooperation, especially to the poorest countries and those that experienced crisis, such as violent conflicts and natural disasters. Japan has facilitated South-South cooperation and triangular cooperation, and these should be expanded.

Japan has been actively engaged in the preparation of the 2030 Agenda. I expect Japan to continue to be active in the implementation phase, taking a leadership role in the international community. I am also looking forward to Japan's voluntary national review that will be prepared this year.

(3) Promote engagement of youth, women, and the private sector, both within Japan and internationally: On the next point, I would like to see more, and many more Japanese private sector firms get involved in the SDGs, both domestically and abroad. The SDGs present an opportunity for private firms to develop and implement business-led solutions and technologies to address sustainable development challenges. Private companies can contribute through their core business activities.

I referred to the power of the youth earlier. The young people of Japan can and should be more actively involved in the implementation of the SDGs. Similarly, as I already mentioned, women's role in society should be promoted, including their engagement in the SDGs. I do not mean to single out just these three groups, and leave out others. I just feel that these three groups have a huge potential to contribute much more to the

SDGs implementation.

7 Final remarks

Now that we are in the second year of the 2030 Agenda, we must accelerate actions. There are many players at every level in all parts of the world. It is important that these players collaborate, accelerate joint learning, promote integrated approaches and actions towards achieving the SDGs.

This is a time of immense opportunity, and we must seize the opportunities. Your active engagement, actions, and contributions are so important to make our world a better place for all its people. With your involvement, and working together with others in Japan and other parts of the world, I am certain that we will move with great strength to achieve the 2030 Agenda and the SDGs.

So, ACT NOW!! Now is the time!

Thank you very much.

i The SDGs global indicator framework that was adopted by the UN General Assembly in July 2017 contains 230 indicators.
ii The global indicator framework developed by the Inter-Agency and Expert Group on SDG Indicators (IAEG-SDGs) was

agreed by the UN Statistical Commission held in March 2017. It was subsequently adopted by the Economic and Social Council (ECOSOC) in June and by the UN General Assembly in July of the same year. As such, the SDG global indicator framework containing 230 indicators for the 17 goals were determined.

iii Japan's SDGs Implementing Guiding Principles: http://www.kantei.go.jp/jp/singi/sdgs/dai2/siryou1e.pdf

第2部 持続可能な地球社会をめざして

1 持続可能な開発目標（SDGs）と
グローバル・ガバナンス

滝澤美佐子

1 SDGs 実現の前提としての
グローバル・ガバナンス

2015年9月、「我々の世界を変革する：持続可能な開発のための2030アジェンダ」（Transforming Our World: The 2030 Agenda for Sustainable Development）が国連総会で採択された[1]。その中に、17の目標と169のターゲットから成る持続可能な開発目標（Sustainable Development Goals: SDGs）が示されている。SDGs は基本的に貧困撲滅を主眼とし、持続可能な開発を加速するための国際目標である。しかし、内容は、隣接する地球規模課題を射程にいれた「統合的」なものである。とくに、持続可能な開発と密接に関係する3側面として経済、社会、環境を調和させることを強調し、持続可能な開発のための必要条件として、平和、人権をも統合する。SDGs は、全ての国や全ての社会・人々に適用され実現を目指す普遍的目標である。脆弱な立場に追い込まれている人々を置き去りにしないだけでなく、その必要に特別の焦点をあて優先課題とする。

上記2030アジェンダの文書では、この普遍的、変革的なSDGs17目標のために、「すべての国、ステークホルダーは、協同的なパートナーシップの下、この計画を実行する」(前文2パラグラフ) として多様なアクターによるグローバル・パートナーシップの再活性化を強調する。これは、SDGsに示されたあるべき国際社会像を共有し、共同で実現しようとする試みである。SDGsに示された地球と人類の課題に協働で対処する「地球規模の統治」(グローバル・ガバナンス)[2]がどのように運営されるかが、SDGsの成否を決める。

SDGsの17目標が市民社会、ステークホルダーの関与を得て2年以上にわたって公開コンサルテーションを経て成立したことは、すでにこの目標が多様なアクターに共有される契機となり、公的セクターのみならず民間セクターにおいても普及が加速されている。しかし、SDGsを誰もが知るようになっても、それは2030年までの実施を保証するものでない。SDGsの実効性をどのようなやり方で高めるかが問われなければならない。

2　SDGsの実効性とグローバル・ガバナンスの方法

では2030アジェンダはSDGsの実効性をどのように高めようとしているのだろうか。

SDGsは、ミレニアム開発目標 (Millennium Development Goals: MDGs) もそうであったように、国際条約の形をとらず、努力義務の規範である。その実効性を高めるために、いくつか

の実施手段を提示している。

(1) 関連する条約の実施

まず、SDGsは単独で存在する規範ではない。MDGsの未達成分を引き継いでいるし、持続可能性の側面では、各種の環境条約の達成とも関係してくる。SDGsは基本的に持続的な開発を強化促進して貧困削減を最重要の目標としているが、その方向性は包摂的なものである。開発のみならず隣接分野の重要な国際条約、国際約束等と組み合わせて実現を図ることを前提としている。

2030アジェンダは、宣言10項においてSDGs達成を支える主要原則を明記し、国連憲章等国際法の原則、人権分野では世界人権宣言と発展の権利宣言や人権諸条約、さらにミレニアム宣言、世界首脳会議成果文書 (2005年) を掲げる。開発分野では、MDGsとアディス・アベバ行動目標が不可分な部分とされ、アフリカ連合などの地域的に形成された開発の各種行動計画やイニシアティブも明記されている。環境では気候変動枠組条約 (United Nations Framework Convention on Climate Change: UNFCCC) が目標13の気候変動に関連して「気候変動への世界的対応について交渉を行う基本的な国際的、政府間対話の場である」とする。人権では上記宣言等に加えて締約国の多い条約である子どもの権利条約の遵守やジェンダーの主流化がうたわれている。それ以外にも、SDGsは平和、保健衛生、労働、環境等多くの分野の多国間条約等の積み重ねの上にできたものである。他の関連する現在および将来の国際合意の実現

がすなわち SDGs の実効性を高めるということが想定されている。

(2) SDGs の実施主体

2030 アジェンダ 39 項は、SDGs の実施の精神を「世界的連帯」に求めている。そして、実施の主体として、政府、民間セクター、市民社会、国連機関を含むあらゆる主体を上げ、地球規模レベルで集中的な取り組みを促進する。「グローバル・パートナーシップ」、別言すれば多様な主体による協働[3]による SDGs の実施である。

国家は、自国において SDGs 実現に第一義的責任を負うとするが、政府、地方自治体等が含まれる。また、国家は能力の不足する他国に SDGs 実現の支援をする役割もある。国家は、法的拘束力のない SDGs をローカルアジェンダなどに国内目標化すること、開発目標に取り込むことで、国内実施や国際実施を具体的に促進することができる。

国連システム諸機関、国連専門機関を含めた国際機構の役割も重要である。国連システムでは、とくに国連開発関連機関の首尾一貫した対応を明記している（46 項）。世界銀行やアジア開発銀行などの国際金融機関の役割を、殊にアフリカ諸国、低開発国、内陸開発途上国等脆弱な国々への支援の文脈で明記している（44 項）。さらに、地域的国際機構、地域経済統合など地域レベルでの SDGs の実施（21 項）が、国レベルの実施を後押しすることを期待している。

非国家、民間部門の重視も SDGs 実施の方法の大きな特徴で

あろう。そこでは市民社会組織であるNGO／NPO、協同組合や慈善団体、ビジネスセクターでは小規模企業から多国籍企業まで、他にも教育、保健医療、さらには市民社会一般の我々も実施の主体である。

　SDGsの包括的で統合的な目標について、若者も含むすべての人が実施の責任を担うという、グローバルな秩序像を提示している。

(3)　資金と技術の支援

　SDGsは、開発援助全般と同様、方法として資金援助、技術支援が実現のためには必須の条件になる。各国が自国の財源をSDGs達成に向けるのはもちろんのこと、SDGs達成が困難な国や地域に対して、資金や技術上の支援の措置を行わなければならない。

　2030アジェンダは政府開発援助（Official Development Assistance: ODA）などの国際的公的資金については、民間資金の触媒ともなると位置づけ先進国の責務を明記している。SDGs目標17「持続可能な開発のための実施手段を強化し、グローバル・パートナーシップを活性化する」の資金の項がそれである。MDGs目標8で達成がされていなかったODA供与国の国民総所得（Gross National Income: GNI）比の0.7%を開発途上国に、GNI比の0.15-0.2%を低開発国に提供するよう再度明記している（43項）。

(4) 指標の設定とフォローアップ活動の実施

上記SDGs目標17では、SDGsは実効性を高めるため、資金の問題に加えてフォローアップの仕組みを用意した。

SDGsは各目標に具体的なターゲットをおき、多様な主体が目標を達成に向ける際の行動基準を示した。さらに、2017年7月にはSDGsのフォローアップ活動を支援するために、国連総会が232の指標を採択した。ターゲットと指標は、SDGsのモニタリングと評価のためには必須である。その他、SDGs進捗状況の測定のためには、統計が必要であり、SDGsに合わせたデータの収集も必要である。

指標の設定とともに、SDGsには次の3つのフォローアップ活動が付与された。

第1は各国のレベルでの活動で、国内行動計画（National Action Plan: NAP）の作成と国内レビューである。日本政府の場合は内閣官房に設置された持続可能な開発目標（SDGs）推進本部が、国としてNAPを作成し、実施に関わっている。第2に、国連事務総長による持続可能な開発目標報告書の公表である。国連事務総長は、2016年に初めての持続可能な開発目標報告書を発行した[4]。第3に4年に1度の専門家による持続可能な開発グローバル報告書が予定されている。

3 SDGsの達成とグローバル・ガバナンスの課題

以上、①既存の条約や制度をSDGsと関連づけること、②多

様な主体による協調、③資金と技術の支援、④各政府と国連、専門家によるフォローアップ活動が、SDGs の達成のために 2030 アジェンダが用意した方法である。それは、法的な拘束力や上からの強制力を用いた方法ではない。むしろ全ての主体を包み込んで責任と自覚を求める「大きな枠組み」を決めるものである。2030 年に向けた望ましい国際秩序を掲げ、それを多種多様な主体が自主、自律的に行動し、様々なネットワークで協働をする。そのための資金や技術の移転、進捗状況を確認しながら進めていく。平和構築、人権保障の制度や活動が SDGs を関連づけながら SDGs 目標達成に寄与することが期待される。このことは、SDGs が 15 年間持続し集中的な目標であり続けるための仕掛けである。この大きな枠組みのもとで、SDGs は、多様な主体による自律的な行動による達成方法をも認めるものである。SDGs 実現のグローバル・ガバナンスは従って、上記の①から④に限られない開放的(オープンエンド)な行動を含むものになる。

　この開放されたグローバル・ガバナンスの方法は、重層的に運営されていく「グローバル・パートナーシップ」である。しかし 2030 アジェンダは具体的方法をそれほど詳細に明示していない。SDGs の各目標がどのようにすれば持続し、衝突的・相殺的にならず、実効性を持ちうるかについても具体的には語られず、現場に任されている。

　2030 アジェンダには、「各国の現実、能力及び発展段階の違いを考慮に入れ、各国の政策、優先度を尊重して」という言い回しが随所にみられる。普遍的目標でありながら、実際の実施

には、対処能力による限界を計算に入れてかからなければならない。

　SDGsのこの「大きな枠組み」としての力を梃に、SDGsの目標と下位のターゲットにおいて、運営の中核となる条約や制度、担い手となるアクターがすでに存在している。SDGsの実効性を高めるための専門的知識、技術、統計的データさらに多様な主体を結ぶ有効なネットワークも必要である。今後はベストプラクティスからも学ばなければならない。私たち一人ひとりのSDGsに関わるローカルな行動もSDGs達成への大きな乗数となる。

　SDGsの達成の持続力と多様な主体の相乗効果を生むため、「大きな枠組み」を維持し達成までを伴走する国連の役割は要石である。SDGsの実効性を見守ることは、国連システム以降成立した国際共同体のための国際法が総体として人間らしい生活と地球の存続を約束するか、そう人類が行動できるか、その集大成ともいえそうである。

注

1　国連総会決議A/RES/70/1（2015年9月25日第70回国連総会採択）本文引用の日本語は外務省仮訳（A/RES/70/L1）を用いた。

2　グローバル・ガバナンスの和訳については、グローバルガバナンス学会編大屋根聡・菅英輝・松井康浩責任編集『グローバル・ガバナンス学I　理論・歴史・規範』法律文化社、2018を参照したが、本来は「舵取り」という意味であると説明される。グロー

バル・ガバナンスについては遠藤乾編『グローバル・ガヴァナンスの最前線─現在と過去の間』東信堂、2008 年、吉川元・首藤もと子・六鹿茂夫・望月康恵編『グローバル・ガヴァナンス論』法律文化社、2014 年、なども参照。
3　日本国際連合学会編『多国間主義の展開』国際書院、2017 年、序、p.12 を参照。
4　2016 年、2017 年に続き、持続可能な開発目標報告書 2018 も発表され国連ホームページより以下に入手可能である。https://unstats.un.org/sdgs/report/2018/

2 地球環境の変化とレジリエンスの観点から

吉高神　明

はじめに

2015年9月に開催された国連の持続可能な開発サミットにおいて持続可能な開発目標（Sustainable Development Goals: SDGs）が策定された。これは、2000年の国連ミレニアム・サミットで採択されたミレニアム開発目標（Millennium Development Goals: MDGs）を継承するものであり、2030年までに国際社会が達成すべき17の目標（貧困、飢餓、保健、教育、ジェンダー、水・衛生、エネルギー、成長・雇用、イノベーション、不平等、都市、生産・消費、気候変動、海洋資源、陸上資源、平和、実施手段）を定め、その実現に向けて169のターゲットが設定されている。

国連の持続可能な開発サミットが開催された2015年は、気候変動や防災に関して重要な文書が採択された年であった。

同年3月には、国連防災世界会議において、従来までの「兵庫行動枠組」を継承する国際防災指針として「仙台防災枠組2015-2030」が採択されている。同枠組みでは、災害リスク削

減に向けた4つの優先行動と7つのターゲットが設定されている。

また、同年11月から12月にかけて開催された国連気候変動枠組条約第21回締約国会議（The 21st Conference of the Parties to the United Nations Framework Convention on Climate Change: COP21）では、京都議定書後の2020年以降の地球温暖化対策の国際合意である「パリ協定」が成立している。同協定では、締約国の温室効果ガス削減への取り組みとして「緩和」、気候変動の悪影響への対応としての「適応」に関して詳細に規定されている。

仙台防災枠組、SDGs、パリ協定それぞれには、人間の安全保障の観点から防災の視点が強く反映されている。それは、「災害は、人命や物理的な損失をもたらすのみならず、個々人、特に貧しい人々の生存、尊厳、生活基盤、そしてこれまで達成された開発の成果に大きな影響を与え、社会の持続的な発展を阻むもの」[1]であるからに他ならない。

1 SDGs、気候変動、レジリエンス

「レジリエンス（resilience）」[2]という概念は当初は物理学、心理学、教育学等の分野で用いられていたが、近年、防災分野で広く用いられている。同分野では、災害を引き起こす要因となる現象、物質、人間の活動・状況である「ハザード（hazard）」、ハザード地帯で損害を被る可能性を有する人々、財産、システムとしての「暴露（exposure）」、主体の有する「対応能力（coping capacity)」の関係で「災害（disaster）」が発生すると考える。すなわち、「ハザード」の「暴露」への悪影響が主体の「対応能力」を超えている状況を「災害」に対する「脆弱性（Vulnerability）」としてとらえ、地域、国、都市、コミュニティ、個人等の「レジリエンス」を強化することによって、「脆弱性」の除去・低減を目指すものである。[3] この場合、レジリエンスの強化は、災害の発生前のハザードやリスクに対する予防（被害抑止・軽減）、災害発生中の被害拡大防止（応急・緊急対応）、及び災害発生後の復旧・復興の３つの文脈に位置づけることができる。[4]

　SDGsの17の目標と169のターゲットには、レジリエンスという用語が頻繁に用いられている。目標１の「貧困削減」に関するターゲット1.5は「2030年までに、貧困者や脆弱な人々のレジリエンスを構築し、気候変動関連の極端な事象、その他の経済、社会、環境上のショックや災害に関する暴露や脆弱性を軽減する」としている。また、目標11「包括的、安全、レジリエントで持続可能な都市及び人間居住の実現」では、仙台防災枠組2015 - 2030に沿って、全てのレベルにおける総合的な災害リスク管理の策定と実施を行うべきとされている（目標11.b）。それ以外にも「レジリエントなインフラ構築」（目標9）、「レジリエントな農業の実施」（目標2.4）、「持続可能でレジリエントな建築物の整備支援」（目標11.c）、「全ての国々の気候関連災害や自然災害に対するレジリエンス及び適応能力の強化」（目標13.1）、「レジリエンスの強化などによる海洋及び沿岸の生態系の持続的な管理と保護」（目標14.2）等、気候変動

図1：SDGs、仙台防災枠組、パリ協定との関連性

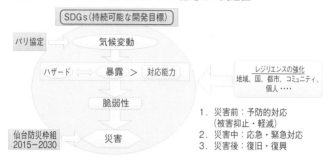

や防災分野での取り組みの重要性が規定されている。

なお、気候変動と防災の観点からSDGs、仙台防災枠組、パリ協定の関係を整理すれば、以下のようになるであろう。（図1参照）。

2 エッセイの紹介

2017年3月に開催されたUNU/jfUNUジュニアフェローシンポジウム『持続可能な地球社会を目指して－私のSDGsへの取組み』の第2セッション「地球環境の変化とレジリエンス」に関しては、8名の方がエッセイを提出してくれた。このうち、当日のセッションに登壇してくれたのは、以下の4名である。

Schulzはレジリエンスの観点から都市の建設資材製造過程における技術革新の必要性を提唱し、資源利用の効率性、循環型経済の確立、地方のエネルギー・アクセス、土地再生・植林事業のあり方等について問題提起を行っている。

Dziubaniukは企業倫理に焦点を当てつつ「持続可能なビジネス」の必要性を主張し、企業（とりわけ、中小企業）に求められる倫理・行動基準やマーケティング上の発想法などについて述べている。

　Hussainは洪水、地震、干ばつなどの自然災害に脆弱な自国のパキスタンを取り上げ、地方政府とコミュニティレベルの災害準備能力強化に向けた方策について論じている。

　Sifuentesはチチカカ湖のトラウト養殖事業に従事する人々の事例を取り上げ、家族、地域団体、NGO、地方政府等の関与を通じて気候変動に対する地域レジリエンス強化への提言を行っている。

　また、当日のセッションには登壇しなかったものの、今回4名の方がエッセイを投稿してくれた。

　Azimはバングラデシュの皮産業を取り上げ、環境及び労働者の健康への悪影響を削減するために自ら立ち上げた「Poli」の取り組みについて紹介している。Botchweyはガーナにおける都市と地方の教育格差との関連で教員の果たすべき役割について論じている。Khanは気候変動の影響が著しい自国のパキスタンで自ら取り組んでいるエネルギー使用法の転換プロジェクトについて紹介している。最後に、洪水早期警戒に向けた天候予測を専門に学んだVuillaumeは、自らの経験を民間セクターでいかにして生かすべきかについて述べている。

注

1　JICA、2009年2月、『課題別指針：防災』　3頁。http://gwweb.

jica.go.jp/km/FSubject0301.nsf/B9EBD9A793E2456249256FCE001DF569/3D09329C78B55A5D4925741700254FC1?OpenDocument（2018 年 2 月 5 日）
2　レジリエンスについては、以下の文献を参照のこと。GSDRC, University of Birmingham, 2014, *Disaster Resilience: Topic Guide*, Birmingham, UK. http://www.gsdrc.org/topic-guides/disaster-resilience/（2018 年 2 月 5 日）
3　防災・減災の文脈における「ハザード」、「暴露」、「脆弱性」、「レジリエンス」等の諸概念については、以下の文献を参照のこと。
ISDR, 2009, *UNISDR Terminology on Disaster Risk Reduction 2009*, United Nations http://www.unisdr.org/we/inform/publications/7817（2018 年 2 月 5 日）
京都大学・NTT リジリエンス共同研究グループ、2012 年、『しなやかな社会への試練：東日本大震災を乗り越える』 日経 BP コンサルティング
4　JICA、2009 年 2 月、『課題別指針：防災』 3-5 頁。

参考文献

1　JICA、2009 年、『課題別指針：防災』 http://gwweb.jica.go.jp/km/FSubject0301.nsf/B9EBD9A793E2456249256FCE001DF569/3D09329C78B55A5D4925741700254FC1?OpenDocument（2018 年 2 月 5 日）

2　ISDR, 2009, *UNISDR Terminology on Disaster Risk Reduction 2009*, 2009, United Nations http://www.unisdr.org/we/inform/publications/7817（2018 年 2 月 5 日）

3　GSDRC, University of Birmingham,2014, *Disaster Resilience: Topic Guide*, 2014, Birmingham, UK. http://www.gsdrc.org/

topic-guides/disaster-resilience/（2018 年 2 月 5 日）
4 京都大学・NTT リジリエンス共同研究グループ、2012 年、『しなやかな社会への試練：東日本大震災を乗り越える』 日経 BP コンサルティング

3 持続可能な開発目標（SDGs）と
グローバル・シティズンシップ：
持続可能な社会を支える人々の協力・協働・共生

杉 村 美 紀

　ミレニアム開発目標（Millennium Development Goals: MDGs）に代わり、国連が定めた新たな目標、持続可能な開発目標（Sustainable Development Goals: SDGs）は、2012年から3年にわたって検討され、2015年8月に合意文書「我々の世界を変革する：持続可能な開発のための2030年アジェンダ」[1]のなかにまとめられた。SDGsには17の目標と169のターゲットが掲げられているが、そのなかにあって「持続可能な社会の担い手」と関連して取り上げられているのが「グローバル・シティズンシップ」という概念である。「2030年アジェンダ」には、その「前文」において「すべての国及びすべてのステークホルダーは、協同的なパートナーシップの下、この計画を実行する。」とあり、「宣言」の第52項には、「人々の、人々による、人々のためのアジェンダであり、そのことこそが、このアジェンダを成功に導くと信じる。」と述べられている。MDGsが途上国社会を中心とした目標設定であったのに対し、「誰も置き去り

にしない」ことを掲げる SDGs は先進国を含むすべての国や社会を対象としており、グローバル・シティズンシップはまさにそれを実現するための担い手を象徴する概念ということができる。

　グローバル・シティズンシップという概念が登場するのは、SDGs の「目標4：すべての人に包摂的かつ公正な質の高い教育を確保し、生涯学習の機会を促進する」の中の「ターゲット4.7」である。同ターゲットには、「2030年までに、持続可能な開発のための教育及び持続可能なライフスタイル、人権、男女の平等、平和及び非暴力的文化の推進、グローバル・シティズンシップ、文化多様性と文化の持続可能な開発への貢献の理解の教育を通して、全ての学習者が、持続可能な開発を促進するために必要な知識及び技能を習得できるようにする。」と述べられている。この概念の定義については様々な解釈があるものの、1990年に発表された「万人のための教育（Education for All: EFA）」の取り組みや、2000年のダカール枠組みにおける EFA の再確認、さらに MDGs とともに、国際社会が取り組んできた「持続可能な開発のための教育（Education for Sustainable Development: ESD）」を通じて培われてきた理論や実践は、SDGs についての今後の組り組みの大きな礎となるものといえる[2]。「2030年アジェンダ」の発表に先立ってユネスコ（国際連合教育科学文化機関、United Nations Educational, Scientific and Cultural Organization: UNESCO）が2015年5月に招集した世界教育フォーラム（World Education Forum: WEF）でも、EFA のゴール及び MDGs の

進捗や教訓を評価したうえで、SDGs でうたわれている「変革と普遍」は EFA や MDGs の「未完の案件」に取り組むことを意味していると述べられている[3]。

そもそもグローバル・シティズンシップが重視されるべき背景には、今日的課題に向き合い、将来に向けて課題解決にあたる「市民」という存在が、社会の多様化や多文化化が進むなかで、従来以上に重要になっていることが挙げられる。「シティズンシップ」が示す「市民性」には、人間の尊厳を重視し、個人の権利と責任を認識したうえで現実の社会と向き合うことや課題解決に向けて取り組むこと、ならびに異文化間の差異と多様な文化的背景を持つ他者を理解し、かつ協働できる力をいかに習得するかということが求められている。特に国境を越える人々の移動が活発化している今日にあっては、当該国の国籍をもった国民だけではなく、難民や移民、外国人労働者などの国籍を持たない人々や、国際結婚や帰化によって国籍は持っていても異なる文化的背景を持っている人々など、様々な立ち位置にいる住民が共存する[4]。そうした社会にあっては、人々の協働・協力があってはじめて共生が可能となる。持続可能な社会の担い手としてのグローバル・シティズンシップが重視されるのもこのためである。

今日では、そうした「市民性」を育てる教育として、「グローバル・シティズンシップ教育（Global Citizenship Education: GCED）」の役割が期待されるようになっている[5]。すでにこれまでにも、「持続可能な開発のための教育（Education for Sustainable Development: ESD）が国際社会で広く展開されて

きたが、SDGs の場合には、GCED や ESD を通じて「ローカル及びグローバルな課題を解決することのできる、能力、価値観や姿勢を発達させることができる」[6]とされており、国家の枠組みを越えた国際社会を対象とすることもあれば、居住社会における身近な課題に取り組む場合もある。しかもそこでの活動主体は、従来のように政府や国際機関、国連システム、民間組織、市民団体といった組織だけではなく、個人にも焦点を当てている点に特徴がある。

　そして実際に様々な教育活動が展開されるようになっていることは、グローバル・シティズンシップ教育の展開を考える上でも大変興味深い。例えば、世界180カ国以上の地域に1万校以上あるといわれるユネスコスクールの活動はそのひとつである。ユネスコスクールはユネスコ憲章に示されたユネスコの理念を実現するため、ユネスコスクールのネットワークを活用した世界中の学校と生徒間・教師間の交流や、地球規模の諸問題に対処するための新しい教育内容や手法の開発、発展を目指している。しかしながら、必ずしも国際的な視野だけではなく、むしろそれぞれの学校がある地域社会とのローカルな連携を図りながら活動を展開している例もみられる。日本では、「国連持続可能な開発のための教育の10年（United Nations Decade of Education for Sustainable Development: DESD）」が始まった2005年からユネスコスクールの数が急増し、2017年10月時点で1034校と1カ国あたりの加盟校数では世界最多となっており、ESD の拠点として自分たちの住む地域社会を拠点とした学びが構築されている[7]。

また各国や地域における教育の国際化の進展に伴い、留学生移動が活発化するとともに国境を越える多様な国際共同プログラムが展開されるようになっていることもグローバル・シティズンシップ育成と関連がある。こうしたプログラムには、各教育機関の連携協定によるものの他、近年では政府間の協力や、欧州連合（European Union: EU）や東南アジア諸国連合（Association of South East Asian Nations: ASEAN）など地域機構が実施するプログラムなど多様である[8]。これらは、国連大学（United Nations University: UNU）が展開している大学院教育やグローバル・セミナーとともに、多様な文化的背景と多様な国籍をもった学習者が、それぞれの政治体制や宗教、言語等の違いを超えて学ぶアカデミックなプラットフォームであり、地球規模の課題に対して学際的に協働してアプローチする場となっている。

同時にこうした多様な教育実践の場は、その定義が必ずしも明確にはされていないグローバル・シティズンシップの意味をその時々のプログラムやそこに参加する人々の交流を通じて問い直す機会でもある。入江昭（1998）は、国際関係における国際主義の潮流に着目し、「『他者』を疑い嫌悪する自国中心主義を克服し、相互依存や協調が進みあらゆる国を受け容れる国際共同体の樹立に向けて、自国の対外政策や国際政策を再考しようとする」[9]ことの重要性を指摘し、その際に国と国との文化的なつながりに焦点をあてた「文化的国際主義（Cultural Internationalism）」という考え方を提示している。そして「文化面での国際関係を担ってきたのは、しばしば国家というより

は個人や民間の団体だったということ」[10] を示唆している。グローバル・シティズンシップはまさに、そうした文化的国際主義の立場にたつことによって国際文化交流を支える主体となるものであり、グローバルな問題から生活と結びついたミクロな実践にいたる様々な次元で、協力や協働を重ねることで、持続可能な共生社会の実現に寄与する存在であるということができる。

注

1 「我々の世界を変革する：持続可能な開発のための 2030 アジェンダ（A/70/L.1）」http://www.mofa.go.jp/mofaj/files/000101402.pdf（2018 年 2 月 5 日閲覧）

2 *Education for All Movement* Retrieved February 7, 2018 from http://www.unesco.org/new/en/education/themes/leading-the-international-agenda/education-for-all/

3 「『仁川（インチョン）宣言』2030 年に向けた教育：包括的かつ公平な質の高い教育及び万人のための生涯学習に向けて」（世界教育フォーラム 2015、2015 年 5 月 21 日）http://www.mext.go.jp/unesco/002/006/001/shiryo/attach/1360521.htm（2018 年 2 月 6 日閲覧）

英文は *Education 2030 Incheon Declaration: Towards inclusive and equitable quality education and lifelong education for all* Retrieved February 6, 2018 from

http://uis.unesco.org/sites/default/files/documents/education-2030-incheon-framework-for-action-implementation-of-sdg4-2016-en_2.pdf

4 杉村美紀 (2015)「ヒトの国際移動と『グローバル・シティズンシップ』日本異文化間教育学会編『異文化間教育』42 号、2015 年 8 月、30‒44 頁。

5 田中治彦・杉村美紀編著 (2014)『多文化共生社会における ESD・市民教育』上智大学出版

6 前掲「『仁川宣言』2030 年に向けた教育」第 9 項目 (ibid. *Education 2030 Incheon Declaration.* Section 9.p.8.)

7 日本ユネスコ国内委員会「ユネスコスクール」ウエブサイト。
http://www.mext.go.jp/unesco/004/1339976.htm （2018 年 2 月 7 日閲覧）
ユネスコ・アジア文化センター「ユネスコスクールへようこそ！」ウエブサイト。
http://www.unesco-school.mext.go.jp/aspnet/ （2018 年 2 月 7 日閲覧）

8 杉村美紀 (2013)「アジアの高等教育における地域連携ネットワークの構造と機能」『上智大学教育学論集』47 号、2013 年 3 月、21‒34 頁。

9 入江昭（篠原初枝訳）(1998)『権力政治を超えて―文化国際主義と世界秩序―』岩波書店、21 頁。原書は、Iriye, A. (1997) *Cultural Internationalism and World Order.* The Johns Hopkins University Press.

10 入江昭（篠原初枝訳）(1998) 同上書、236 頁。

4 持続可能な開発目標（SDGs）達成に向けた国連大学の取組みと今後の展望

齊藤　修

1 国連大学全体での取組み

　1975年以来、国連大学（United Nations University: UNU）は実社会の問題に取り組むための学術的根拠に基づく研究活動を行ってきた。UNUの研究の目的は、次の世代を含めた人類が直面する地球規模の諸問題解決を目指し、信頼性と客観性の高い助言を国連加盟国や関係機関等に提供することである。

　持続可能な開発目標（Sustainable Development Goals: SDGs）が採択された2015年以降は、「UNU Strategic Plan 2015-2019（国連大学戦略プラン2015-2019年）」のもとで組織の運営が行われている。この戦略プランでは、以下の三つのテーマ領域が提示されている。

- 平和とガバナンス
- 世界の開発と一体性
- 環境、気候、エネルギー

UNUの研究は、17項目のSDGsすべてに関わっており、世

界中から集まった 400 人以上の研究者が、180 を超える数のプロジェクトに従事し、SDGs に関連する研究を進めている。2017 年からは「国連大学と知る SDGs-Sustainable Development Explorer」と銘打ったキャンペーンを展開し、SDGs のそれぞれの目標別に、UNU に所属する研究者たちとその研究内容をわかりやすく紹介している (https://jp.unu.edu/explore)。

2　国連大学サステイナビリティ高等研究所 (UNU-IAS) での取組み

　東京を拠点とする国連大学サステイナビリティ高等研究所 (United Nations University Institute for the Advanced Study of Sustainability: UNU-IAS) の使命は、サステイナビリティとその社会的・経済的・環境的側面に注目しながら、政策対応型の研究と能力育成を通じて、持続可能な未来の構築に貢献することである。また、気候変動や生物多様性、防災等の分野での国際的な政策決定や国連システム内の議論に貢献することが求められている。そのため、UNU-IAS の研究教育活動には、「持続可能な社会」、「自然資本と生物多様性」、「地球環境の変化とレジリエンス」という 3 つのテーマ領域が設定されている。これらを SDGs と紐づけるなら、「持続可能な社会」は SDGs のすべての目標に関係し、「自然資本と生物多様性」は主に目標 14 (海洋資源・海域生態系) と目標 15 (陸域生態系) に対応し、「地球環境の変化とレジリエンス」は主に目標 9 (レジリエントなインフラ構築) と目標 13 (気候変動) に対応する。

SDGs の個別目標の達成に向けた取り組みを途上国だけでなく、先進国も含めて強化することが重要であるが、その際、しばしば指摘されるのが目標間での相互連関（インターリンケージ）である。すなわち、ある目標の達成が他の目標の達成に悪影響を与えるトレードオフをいかに避け、逆にある目標の達成が他の目標の達成を促すようなシナジーを生みだしていくことが重要になる。このような SDGs 間のインターリンケージについては、国際科学会議（The International Council for Science: ICSU）が『SDGs 間の相互関係に関するガイド：科学から社会実装』と題する包括的なレポートを 2017 年に発表している。このレポートでは、例えば、目標 2（飢餓の克服・持続可能な農業）を起点とした場合、目標 1（貧困撲滅）、目標 3（健康）、目標 6（水資源）、目標 7（エネルギー）、目標 14（海洋資源・海域生態系）と目標 15（陸域生態系）などとの関係が重要であることが指摘されている。こうした SDGs 間のインターリンケージの関係性は、国や地域の置かれた状況や過去からの経緯によって異なることから、現在、UNU-IAS ではアフリカの複数の対象国を対象として、インフラと都市に関する目標 9 と目標 11 の推進をメインとしたとき、SDGs 間のシナジーの最大化とトレードオフの回避を同時追求できる政策オプションを模索する研究に取り組んでいる。言い換えると、アフリカの開発のカギである「都市問題」を中心に複数の目標への貢献を通じたインパクト拡大を実現する政策モデルを提示し、さらに民間企業や自治体、市民社会の参加が継続・拡大するようなプラットフォームを構築することをめざしている。

SDGsではまた、複数の目標にわたって横断的にレジリエンス概念の重要性が言及されている。レジリエンス（回復能力）とは、何らかの撹乱に対して、システムがそれを吸収し、機能や構造を維持する能力を示す概念であり、1973年にC.S.Hollingが生態学の概念として提唱した。その後、生態経済学のグループが社会生態レジリエンスとして展開した。定常的な状態のもとで工学的な最適化や効率化を追及するこれまでの資源管理アプローチに対し、レジリエンスの視点からは、常に動的に変化する状況のなかで社会生態システムの機能をいかに維持するか、そのためのオプションと変化に応じた自己再編が重視される。UNU-IASは、東京大学やガーナ大学等との共同研究で、北部ガーナを対象とした研究プロジェクト（Enhancing Resilience to Climate and Ecosystem Changes in Semi-Arid Africa: An Integrated Approach: CECAR-Africa）の一環として、特に気候・生態系変動に対する統合的なレジリエンス評価のための評価指標開発と農村集落への適用を行った。

　このプロジェクトでは、地域住民の生計・社会経済調査に基づき、地域の歴史や伝統知を尊重した適応策の提案と地域開発、洪水や渇水に対する地域のレジリエンス向上を試みた。具体的には、現地の研究者や政府関係者とともに、地域のレジリエンスを（1）生態学的（農業生態系の多様性、作付け品種の多様性など）、（2）工学的（気象情報の早期警戒システム、土壌管理、集水・貯水技術など）、（3）社会経済的（生業・収入源の多様化、災害システム、防災教育など）な側面から評価する指標群を提案し、それをガーナ北部10カ所の対象集落を評価した。その

評価結果を含めて研究成果を共有しつつ、今後の対策について議論するワークショップを対象集落で開催し、一連のワークショップを通じ、研究成果が地域住民の生計や暮らしの向上や気候・生態系変動への適応にいかに役立つのか、地域の人々と共に考え、行動するためのパートナーシップ構築を進めた。このようなボトムアップでのパートナーシップ構築の取組みは時間と労力を要するが、その一方でプロジェクト単独では想定しえなかった研究や社会実装への展開が促進されるほか、関係者が当事者意識を持って主体的に取組むことが促され、それによってプロジェクトの諸活動の継続性が自ずと担保されるという大きなメリットがある。

　レジリエンスと同様に、SDGs では「誰も置き去りにしない」、いわば包摂性ないし包括性（inclusiveness）がすべての目標にまたがる重要な概念であるが、この包摂性・包括性をそれぞれの国・地域の実情に即して具体的なアクションに落とし込んでいくには、科学的知見をさらに蓄積していく必要がある。例えば、国連大学は、国連環境計画（United Nations Environment Programmes: UNEP）など複数のパートナーと共同で、代表的な経済指標である国民総生産（Gross Domestic Product: GDP）を補完する指標として、包括的富指標（Inclusive Wealth Index: IWI）の概念を提唱している。この新しい指標は、持続可能性に焦点を当て、長期的な人工資本（機械、インフラ等）、人的資本（教育やスキル）、自然資本（土地、森、石油、鉱物等）を含めた、国の資産全体を評価し、数値化している。国連大学は、この包括的富指標で世界各国を評価したレポートを 2012

年から発行しており、2018年には第三次レポートを発表予定である。

　一方、SDGsに貢献する人材育成のプログラムとして、国連大学では、2013年からアフリカでのグローバル人材育成プログラム（Global Leadership Training Programme in Africa: GLTP）を実施している。アフリカの諸課題解決に貢献する人材の育成／国際機関、NGO職員として世界で活躍できるグローバル人材育成を目的として、日本のリーディング大学院およびアフリカ各国の大学や研究機関と連携して、アフリカをフィールドとした大学院生の調査研究をサポートしている。毎年、日本全国の大学院生（修士・博士課程）を10名程度選抜し、アフリカ諸国のパートナー大学への派遣（2〜10か月程度）を通じて、将来のリーダーとしての資質を磨き、研究デザイン、実施から研究成果の社会還元までの一連のプロセスを学んでもらっている。アフリカでの持続可能な開発におけるさまざまな課題や問題の解決に取り組み、その研究成果を日本のみならず、アフリカの現地関係者へフィードバックすることで、日本とアフリカの関係強化とSDGs達成に取り組んでいる。

3　今後の展望

　国連大学では、今後も間違いなくSDGs達成に資する研究教育が強化されることになる。特に、前述したSDGs間のインターリンケージやレジリエンス評価については、研究蓄積が十分とはいえず、まだまだ研究課題が多く残っている。また、包括的

富指標は、現状では国レベルでの適用に留まっており、地域や自治体単位などより小さいスケールで適用するための手法開発が求められている。

　フューチャー・アースは、持続可能な地球社会の実現をめざす国際協働研究プラットフォームであり、「人類が持続可能で公平な地球社会で繁栄する」というビジョンの実現に向け、ダイナミックな地球の理解、地球規模の持続可能な発展、持続可能な地球社会への転換、という3つの大きなテーマで研究を2015年から推進している。フューチャー・アースでは、SDGs達成を革新的なアイディアで推進する世界各地での具体的な取組みをSDG実験室（SDG Labs）として認定している。国連大学もフューチャー・アースのメンバーであり、今後はこのグローバルなネットワークと密に連携した研究教育を展開することを通じて、SDGsに関連するグローバルな科学-政策対話とローカルなアクションの両方の強化に向けて学術的根拠に基づいた貢献を具現化していきたい。

5 持続可能な開発目標(SDGs)とビジネス:
グローバル規範構築の可能性

秋 月 弘 子

　2001年に発表されたミレニアム開発目標(Millennium Development Goals: MDGs)は、おもに途上国の開発課題に取り組む国際開発目標であった。しかしMDGsは、国連や経済協力開発機構(Organisation for Economic Co-operation and Development: OECD)などの国際機構によりトップダウン形式で策定されたものであり、当事者である途上国自身の声が十分に反映されていないという不満があった。

　そこで、2015年9月に採択された国連総会決議70/1「持続可能な開発のための2030アジェンダ」の中で規定された持続可能な開発目標(Sustainable Development Goals: SDGs)は、MDGsの後継の2016年から2030年までの国際目標として、途上国政府のみならず、市民社会組織をも含む参加型アプローチにより2年以上の時間をかけて練り上げられた。

　MDGsが8つの目標を持つ途上国中心の開発目標であったのに対し、SDGsは17の目標を持ち、経済成長、公正で包摂的な社会の構築、地球環境の保護などをめざしている。また、

再生可能性エネルギーの利用拡大や児童虐待の撲滅、商品廃棄物の半減、海洋資源の保護など、先進国や企業などにも当事者としての取り組みを求める、普遍的な目標となっている。

日本政府も、2016年5月、関係機関の連携および効果的なSDGs推進を行うため、内閣に持続可能な開発目標（SDGs）推進本部を設置し[1]、同年12月には「持続可能な開発目標（SDGs）実施指針」[2]を策定し、1.あらゆる人々の活躍の推進、2.国内外における健康・長寿の達成、3.成長市場の創出、地域活性化、科学技術イノベーション、4.質の高いインフラ、強靱な国土の整備、5.省・再生エネルギー、気候変動対策、循環型社会、6.生物多様性、森林、海洋等、環境の保全、7.平和・安全・ガバナンス、8.SDGs実施推進の体制・手段、という8つの優先課題を特定し、人々（People）、繁栄（Prosperity）、地球（Planet）、平和（Peace）、パートナーシップ（Partnership）の5Pに取り組んでいる。さらに、2017年12月には、「SDGsアクションプラン2018」[3]を策定し、官民を挙げたSDGsを推進する経営および投資の促進、SDGsを原動力とした地方の創生、SDGsの担い手としての次世代や女性のエンパワーメントなどに言及している。

また、2020年に開催される東京オリンピック・パラリンピック大会の組織委員会も、大会に必要な物資およびサービスの調達の際に環境・社会・経済の持続可能性にも配慮した調達を行うために、「持続可能性に配慮した調達コード」[4]を策定している。その中で、SDGsを尊重し、地球温暖化や資源の枯渇などの環境問題や人権・労働問題の防止、公正な事業慣行の推進や

地域経済の活性化等への貢献を考慮に入れた調達を実現することを目指している。

　さらに、一般社団法人日本経済団体連合会（経団連）も2017年11月、企業が守るべき指針を記した「企業行動憲章」[5]を改定し、その中で、企業は「持続可能な社会の実現を牽引する役割を担う」と明記した。これにより、企業にとって最も大切な経営理念にSDGsを採り入れることを求め、事業を通じて貧困や環境など地球規模の課題の解決に貢献していくよう呼びかけている。さらにこの企業行動憲章では、国連が2011年に採択し、人権を尊重する企業の責任に言及した「ビジネスと人権に関する指導原則」[6]に従い、これまでの企業行動憲章にはなかった「すべての人々の人権を尊重する経営を行う」という人権条項を盛り込んでいる。今後経団連に所属する企業は、持続可能な社会の実現が企業の発展の基盤であることを認識し、環境・社会・企業統治（Environmental, Social and Corporate Governance: ESG）に配慮した経営の推進による社会的責任への取り組みを進めることになる。

　これらの企業の取り組みを後押しするのが、国連の責任投資原則（Principles for Responsible Investment: PRI）[7]に基づいた投資である。責任投資原則とは、2006年に国連環境計画（United Nations Environment Programmes: UNEP）および国連グローバル・コンパクト（United Nations Global Compact: UNGC）が提唱したイニシアティブであり、機関投資家による投資分析と意思決定のプロセスにESG課題を組み込む、投資対象の企業に対してESG課題を求める、資産運用業界におい

て同原則を実行するよう働きかける、などの6つの原則からなる。これは、ESG課題に取り組む経営が企業を長持ちさせ、結果として長期的な利益につながるという考えに基づいており、2008年のリーマンショック後、目先の利益ばかりを追い求めたことへの反省から、持続可能な社会の構築に向けて貢献できているかどうかに着目して企業を選別する投資手法（ESG投資）である。機関投資家にとっては、ESG課題を考慮する事が投資リスク管理になると同時に、環境と社会全体に利益をもたらし、社会的責任を果たすことになるとされている。このような資金の流れを作ることで、SDGsが目指す環境や貧困問題の解決、公平、公正な社会づくりなどにもつながることが期待される。

現在、ESG投資はとくに欧米の投資家や企業に急速に拡大しており、2016年の投資の中でESG投資が占める割合は、欧州で52.6%、米国で21.6%、日本では3.4%となっている[8]。

これまで述べてきたSDGs、「持続可能性に配慮した調達コード」、「企業行動憲章」、「ビジネスと人権に関する指導原則」、「責任投資原則」などはすべて、国際条約でも、いずれかの国の国内法でもなく、したがって、法的拘束力は有していない。それらの基準を遵守することが期待される、いわばソフト・ローである。

しかし、「持続可能性に配慮した調達コード」を遵守することによりオリンピック・パラリンピック大会において物品およびサービスの提供が可能となったり、「責任投資原則」を遵守することで資金が集まったりするのであれば、そこにビジネス

利益が生まれ、企業にはそれらの基準を遵守するインセンティブが生まれる[9]。そして、実際に企業がSDGsに取り組もうとする機運が生まれつつある。

つまり、ビジネスの世界においては、もはやSDGsに法的拘束力がないことは問題とはならず、法的拘束力がなくても守られるようになるのである[10]。

世界は、国家間関係を中心とした「国際」社会から、企業や市民社会組織などの非国家主体も重要な行為主体とみなされる「グローバル」社会へと変容してきた。その変化にともない、そこで適用される規範も、法的拘束力を有し、おもに国家間の権利義務関係を規律する「国際法」中心の時代から、法的拘束力はないものの、SDGsのような「グローバル」規範が、企業などの非国家主体をも規律していく時代へと変化してきている[11]。

企業によるSDGsの取り組みは、真のグローバル社会におけるグローバル規範の認識の一例と考えられるのではないだろうか。国家だけでなく、地球上のすべての人を対象とするSDGsが、すべての人によって推進され、「誰も置き去りにしない」真のグローバル規範が構築されることが期待される。

注

1 　首相官邸、2016年、「持続可能な開発目標（SDGs）推進本部」、首相官邸。https://www.kantei.go.jp/jp/singi/sdgs/（2018年1月22日）。

2 　首相官邸、2016年、「持続可能な開発目標（SDGs）実施指針」、

首相官邸。https://www.kantei.go.jp/jp/singi/sdgs/dai2/siryou1.pdf（2018年1月22日）。

3　首相官邸、2017年、「SDGsアクションプラン2018〜2019年に日本の『SDGsモデル』の発信を目指して〜」、首相官邸。https://www.kantei.go.jp/jp/singi/sdgs/dai4/siryou1.pdf（2018年1月22日）。

4　東京オリンピック・パラリンピック組織委員会、2017年、「東京2020オリンピック・パラリンピック競技大会　持続可能性に配慮した調達コード（第1版）」、東京オリンピック・パラリンピック組織委員会。https://tokyo2020.jp/jp/games/sustainability/sus-code/wcode-timber/data/sus-procurement-code.pdf（2018年1月22日）。

5　一般社団法人日本経済団体連合会、2017年、「企業行動憲章」、一般財団法人日本経済団体連合会。http://www.keidanren.or.jp/policy/cgcb/charter2017.html（2018年1月22日）。

6　国連広報センター、2011年、「ビジネスと人権に関する指導原則：国際連合「保護、尊重及び救済」枠組実施のために（A/HRC/17/31）」、国連広報センター。http://www.unic.or.jp/texts_audiovisual/resolutions_reports/hr_council/ga_regular_session/3404/（2018年1月22日）。

7　責任投資原則、2006年、「責任投資原則」、責任投資原則。https://www.unpri.org/download_report/18940（2018年1月22日）。

8　北郷美由紀「投資　変わる前提」『朝日新聞』2017年11月18日、朝刊。

9　吾郷も、「サプライチェーン・マネジメントの観点から発出される各種の条件は、ほぼ法的規範（またはそれ以上）のような力

を持ってくることになる。なぜならば、その条件を満足しない限り、仕事が取れない」と指摘している。吾郷真一「CSR―法としての機能とその限界」『季刊労働法』234号、2011年秋季、51ページ。
10　吾郷は、「適当なフォローアップがなされた場合には、元の文書が法的拘束力のないものであっても、結果的には法的重要性が高まる」ことを指摘している。吾郷、同上。
11　吾郷も、SDGsが従来の開発戦略と大きく異なる点として、国家だけでなく民間主体に重要な役割を課している点を指摘している。吾郷真一「持続可能な開発目標（SDGs）と国際労働基準」国際人権法学会編『国際人権』2017年報、第28号、2017年、114ページ。

第3部 「誰も置き去りにしない」社会をめざして

「誰も置き去りにしない」:
持続可能なグローバル社会のための
ガバナンス

勝 間　　靖

はじめに

　持続可能なグローバル社会をつくるために、国連加盟国は、2030年までに持続可能な開発目標（Sustainable Development Goals: SDGs）を達成すると合意した。SDGsの達成へ向けて、国家間および国内における格差を是正する政策をとることで、包摂的な（inclusive）な社会をつくることを目指している。つまり、持続可能なグローバル社会の特徴の一つは、「誰も置き去りにしない（no one will be left behind）」ということである。そして、そのためには、社会的に排除されやすい脆弱な人びとがSDGsの達成へ向けたプロセスに参画できるようにするグローバル・ガバナンスが必要とされている。

　本稿では、最初に、社会的に排除されやすい脆弱な人びととして、難民に注目する。そして、難民を置き去りにしないために、難民の恒久的な解決策を踏まえたうえで、それに至るプロセスにおいて、とくにSDGsの目標3である「すべての人に健

康と福祉を」にいかに取り組むべきかを議論する。次に、社会的排除の例として、新興感染症の脅威にさらされた脆弱な人びとに光を当て、事例としてエボラ出血熱を取り上げる。そして、政府が、国内的に対応できていないのに、国際的な支援を要請しない場合において、国際社会として感染者が置き去りにされないよう保護する責任を果たすためのグローバル・ガバナンスのあり方について議論する。

1　難民を置き去りにしないために

ドイツのコメディ映画『はじめてのおもてなし（Willkommen bei den Hartmanns)』が、国連 UNHCR 難民映画祭 2017 の初日の 2017 年 9 月末に、東京の渋谷で上映された[1]。ミュンヘンの郊外にある閑静な住宅街に住むハルトマン一家が一人の難民を受け入れる。引退を拒む医師である夫リチャードが反対するにもかかわらず、教師を引退した妻アンジェリカが難民の受入れを決めてしまうのだ。難民の青年ディアロの受入れを契機に、家族のなかで、コミュニティのなかで、そして国との関係で波紋が広がっていく。

「難民を受け入れることで具体的に何が起こりうるか？」という本来は深刻な話だが、コメディとして描かれているため、観客は、差別や偏見に悲しむだけでなく、ときには笑いながら、共感を深めた。サイモン・バーホーベン（Simon Verhoeven）監督によって 2016 年にドイツで制作された映画で、2017 年にはドイツ・アカデミー賞観客賞、バイエルン映画祭で作品賞と

プロデューサー賞を受賞している。日本でも、『はじめてのおもてなし』は、2018年1月に東京の銀座で封切りされたのち、各地の映画館で公開された[2]。

難民「問題」の解決として、一般的に次の方法が提言されている。(1) 平和になった母国へ帰国すること、(2) 一時的に避難した周辺国での定住、(3) 第三国定住、の三つである。しかし、これ以外にも、そもそも難民が発生しないように、紛争や深刻な人権侵害を解決していくための国際的な協力が重要である。また、難民を、負担のかかる「問題」として扱わず、新しい活力を生み出す「機会」として捉える視点も必要になってくる。『はじめてのおもてなし』で描かれているのは、解決方法(3)「第三国定住」における課題である。

2016年に国連総会で採択された『難民と移民に関するニューヨーク宣言』[3]に基づき、国連難民高等弁務官事務所（United Nations High Commissioner for Refugees: UNHCR）は「難民に関するグローバル・コンパクト」を2018年に策定できるよう主導している。こうしたなか、「人間の安全保障」の視点から、難民を置き去りにしないために何が必要かを考えたい。

(1) 難民を生み出す暴力と人権侵害

さて、青年ディアロは、ボコ・ハラムによる暴力から逃れるため、ナイジェリアを出て難民となった。映画では、ドイツでの難民の受入れに焦点が絞られており、難民を生むナイジェリアの状況の説明はディアロが小学校で自分の生い立ちを説明する場面に限定される。ここでは、筆者が監修したビデオ『BBC

世界の諸問題と子どもたち〜貧困・紛争・暴力にさらされる子どもの権利を考える』(BBC 2017)[4]の第3巻「ボコ・ハラムに拉致された少女たち」などを参考にして、ボコ・ハラムについて少し解説しながら、解決方法（1）「平和になった母国へ帰国すること」について考えたい。

　アフリカ大陸中央部に位置するチャド湖は、チャド、ニジェール、ナイジェリア、カメルーンの4か国にまたがっている。そのチャド湖流域では、およそ2100万人の人びとが住んでいるが、とくに2013年以降、武力紛争による人道危機が悪化している。推定260万人が家を追われて強制移住させられているが、そのうち140万人が子どもである。強制移住の結果、多くの難民や国内避難民が発生している。とくに、紛争地から逃れられない国内避難民の子どもたちは、暴力、栄養不良、疾病のほか、教育の欠如に苦しめられている。ナイジェリア北部だけでも、2万人の子どもたちが家族から引き裂かれている（UNICEF 2016）。

　ボコ・ハラムという名の武装集団は、2009年以降にナイジェリア北部を拠点として数千人から構成された、イスラム教のスンニ派を名乗る過激派組織である。イスラム教を掲げて活動しているが、その極端で過激な解釈や、それに基づく暴力や深刻な人権侵害は、多くの穏健なイスラム教徒からは受け入れられていない。

　この武装集団の特徴の一つとして、子どもや女性を、強制徴募し、自爆テロやスパイ活動などに悪用することがある。2014年以降、86人の子どもが4か国において自爆テロに使われた

(UNICEF 2016)。もう一つの特徴は、宗教に基づかない世俗教育と、キリスト教徒とを敵視することである。ナイジェリアでは、人口の約5割を占めるイスラム教徒が主に北部に、約4割のキリスト教徒は主に南部に住んでいるほか、全国的に伝統的な固有の宗教がある。

国連人権高等弁務官は、ボコ・ハラムによる一般市民の殺害や拉致、戦闘行為への子どもの関与、性暴力、拷問は、「子どもの権利条約」をはじめとする国際人権法や国際人道法に違反した行為であると報告している（United Nations High Commissioner for Human Rights 2015）。また、国際刑事裁判所は、人道に対する罪と戦争犯罪の申立てに対して、受理許容性を判断するための予備的な検討を進めている。

東アジアにおいても、こうした暴力や深刻な人権侵害による難民が発生している。たとえば、ミャンマーにおけるロヒンギャ民族の難民化は深刻である。西部のラカイン州などに住むロヒンギャ民族の多くは、国籍を与えられず、市民権を剥奪されてきた。さらに、近年には、これまで住んできた土地を追われるような暴力が顕著となってきた。つまり、解決方法（1）「平和になった母国へ帰国すること」が望ましいのは確かであるが、その国における暴力や深刻な人権侵害の構造を変えることは容易でなく、難民が帰国できる条件はすぐには整備できないのである。

(2) 周辺国からの難民を受け入れるヨルダン

ヨルダンは伝統的に難民へ寛容な国であり、周辺国から多く

の難民を受け入れてきた。解決法（2）「一時的に避難した周辺国での定住」に寄与してきた模範的な国だともいえる。古くはパレスチナ難民を、そしてイラク難民を、さらに最近ではシリア難民を数多く受け入れてきた。しかし、近年の多くのシリア難民の流入は、とくに教育や保健医療などの社会サービス費を増大させ、ヨルダン政府の財政を圧迫している。

中東（ヨルダン、レバノン、シリア、ガザ地区、ヨルダン川西岸地区）のパレスチナ難民を支援する国際機関として、国連パレスチナ難民救済事業機関（United Nations Relief and Works Agency for Palestine Refugees in the Near East: UNRWA）がある[5]。UNRWAはおよそ500万人のパレスチナ難民の支援と法的保護のために活動している。その活動分野は多岐にわたるが、四つの人間開発目標として「知識とスキル」「長寿と健康的な生活」「適正な生活水準」「人権」を掲げている[6]。

保健は重要な活動分野の一つであり、310万人のパレスチナ難民がUNRWAの保健サービスの提供を受けている（UNRWA 2017）。UNRWAの保健システムは三つの層から構成されており、①本部では保健政策と戦略を担当し、②その下に運営管理に携わる五つのフィールド保健部があり、③その下にパレスチナ難民に保健サービスを提供する143の保健センターがある。ヨルダンに200万人以上いるパレスチナ難民の歴史は長く、難民キャンプも一見して住宅地に見えるところもある。しかし、近年のシリア人道危機によって、シリアにいたパレスチナ難民がヨルダンを含む周辺国へ再び難民として流入しているため、状況は複雑化している。

人口が約945万人のヨルダンは、すでに200万人以上のパレスチナ難民がいるうえに、さらに70万人弱のシリア難民を受け入れている。シリアとの国境に近いヨルダン北部には、2012年7月にザータリ難民キャンプが設置され、約8万人が支援を受けている。この他にも、いくつかの難民キャンプがあるが、大半のシリア難民は、今では生活の糧を求めて、ヨルダン北部の地方自治体（municipality）のコミュニティに入っているのが現状である。難民キャンプにおける難民への支援に加え、受入れコミュニティ（host community）における難民への支援がこれまで以上に必要とされている（REACH 2014）。

(3) 難民の安全保障へ向けて

　「人間の安全保障」の視点から、難民の安全保障のために何が必要なのだろうか。難民の保護とエンパワーメント（力をつけること）の両面から考えていきたい。ここでは、難民が直面する暴力や深刻な人権侵害などの脅威そのものの軽減を保護と定義し、難民が脅威に対応できるよう強靭性を高めることをエンパワーメントと定義する。

　第1に、映画『はじめてのおもてなし』でみたドイツにおける「第三国定住」は、日本にとっても参考になるだろう。日本はもっと積極的に難民を受け入れるべきだと思うが、差別や偏見など解決して多文化共生の社会をつくると同時に、難民が仕事をして自立的な生活を過ごせるよう、保健サービスや語学を含めた教育支援も重要となる。つまり、現実主義的な難民受入れ計画が必要である。歴史を振り返ると、1975年にベトナム、

ラオス、カンボジアが社会主義体制へ移行したのち、1970年代後半から1980年代をとおして、多くの難民が発生した。その一部は、いわゆるボート・ピープルとして流入し、日本は1万人以上を受け入れた。朝鮮半島の情勢を考えると、近い将来に、同様の状況が発生し得ることは想定しておいた方がよく、現実主義的な難民受入れ計画を策定することが急務だろう。

第2に、難民が「平和になった母国へ帰国すること」ができるよう、暴力や深刻な人権侵害を含む紛争を抱えた国への平和構築支援が重要である。日本は、国際平和協力の推進のため、これまで国連平和維持活動（Peace Keeping Operations: PKO）に積極的に貢献してきたほか、政府開発援助の重点課題の一つとして、平和構築を支援してきた。政府開発援助の実施機関である国際協力機構（Japan International Cooperation Agency: JICA）は、平和構築のための重点課題として、「社会資本の復興に対する支援」「経済活動の復興に対する支援」「国家の統治機能の回復に対する支援」「治安の安定化に対する支援」の四つを掲げている[7]。解決が難しい問題が多いが、難民が直面する暴力や深刻な人権侵害などの脅威そのものを軽減しようとするような、保護につながる活動を今後も拡充していく必要がある。

第3に、難民が「一時的に避難した周辺国での定住」ができるよう、ヨルダンのような難民受入れ国への支援が重要である。日本は、UNHCRやUNRWAへの資金協力をとおして難民の支援と法的保護に貢献してきた。このほか、日本のNGOが難民救済を含めた人道支援により積極的に取り組めるように、

ジャパン・プラットフォームの仕組みをつくってある[8]。最近の特筆すべきこととして、ミャンマーのラカイン州で発生した暴力とそれによるロヒンギャ民族のバングラデシュへの流入に対応して、2017年9月に、日本政府は、国際機関をとおしたミャンマーとバングラデシュでの人道支援を決めた[9]。こうした動きは高く評価できる。

最後に、難民のエンパワーメントのための具体策として、一つの試みを紹介したい。日本では全国的に普及している母子健康手帳であるが、国際的にはあまり一般的でなかったところ、JICAの支援によって国際的に普及してきた（JICA 人間開発部 2012）。インドネシア人医師が関心をもったことを契機に、JICA は1993年からインドネシア版の母子健康手帳の作成を支援することになった[10]。その結果、2006年にはインドネシア全国すべての州で導入されたのである。そして、インドネシアでの経験が、2007年の研修をとおして、さらにパレスチナやアフガニスタンへと伝えられていった。

パレスチナでは、2006年の日本政府からの無償資金協力に基づき[11]、JICAと国連児童基金（United Nations Children's Fund: UNICEF）との協力により、2008年にアラビア語版の母子健康手帳が導入された[12]。その後、UNRWA の事業として、周辺国のパレスチナ難民にも普及していったのである。それが、近年、スマートフォンで使えるように電子化されたことは注目される。

また、これとは別に、パレスチナ難民には生活習慣病で苦しむ者が多いことから、UNRWA は、生活習慣病対策手帳を作

成したところである。これを将来に電子化することができれば、既存の母子健康手帳に新しい生活習慣病対策手帳を加えて、母子だけでなく難民全員を対象とするような、電子版の難民健康手帳という構想も現実味を帯びてくるのではないだろうか。移住することになっても、さらに国境を越えることがあっても、自分の健康データを自分で管理して持ち運ぶことができる。したがって、電子版の難民健康手帳は、どこにいても難民を置き去りにせず、人間の安全保障を向上させることができるツールとして潜在能力が高いと思われる。

2 エボラ出血熱の感染者を置き去りにしないために

次に、急性ウイルス性の新興感染症であるエボラ出血熱の脅威にさらされた人びとに注目する。1976年に現在の南スーダンで発見されたエボラ・ウイルスは、毒性と感染性が非常に高いが、血液や体液に触れなければ周りの人に感染しない。アフリカで10回ほど流行したことがあるが、大規模な流行は2014年まで起こらなかった。それなのに、なぜ、2014年、エボラ出血熱は、西アフリカで多くの命を奪ったのか？ 2016年1月までに、2万8千人以上の感染者、1万1千人以上の死亡者を出さなくてはならなかったのは、なぜか？

以下では、まず、西アフリカにおける感染拡大とそれへの対応の経緯を振り返る。そして、次の二つの研究問題に答えていく。(1) なぜ世界保健機構（World Health Organization: WHO）は「国際的に懸念される公衆衛生上の緊急事態（Public Health

Emergency of International Concern: PHEIC)」を宣言するのに遅れたか？ (2) なぜ国連事務総長は国連エボラ緊急対応ミッション（United Nations Mission for Ebola Emergency Response: UNMEER）を設置しなくてはならなかったのか？

そのうえで、ギニアの事例から得られる教訓をもとに、新興感染症の脅威にさらされた人びとを置き去りにしないためにグローバルヘルス・ガバナンスへ向けて、提案をおこないたい。

(1) エボラ出血熱の感染拡大と国際的対応

2013年12月、隣国のシェラレオネやリベリアとの国境に近いギニア南東部にある森林地帯の都市で、男の子が亡くなった。ギニアにおけるエボラ出血熱の最初の感染者とみられている。2014年3月、ラボでの検査に基づき、ギニアにおけるエボラ出血熱の発生が確認され、ギニア保健省はエボラ流行を宣言した。同月、国際NGOである国境なき医師団は、「地理的な広がりは前例がない（unprecedented）」と国際社会へ向けて警告を発し（MSF 2015）、ギニアとリベリアにおける活動を拡大した。

しかし、5月にジュネーブで開催されたWHO総会では、西アフリカにおけるエボラ出血熱の流行に関して、ギニア保健大臣もWHO事務局長も十分に注意喚起しなかったため、国際的対応は拡充されなかった。

6月になって、危機感を一層高めた国境なき医師団は、「もはや制御できない（uncontrollable）」とし、国際的対応の拡充を求めた（MSF 2015）。また同月、地球規模感染症に対する警

戒と対応ネットワーク (Global Outbreak Alert and Response Network: GOARN) 運営委員会は、WHO に対してより強力な指導力を求めた。

これを受けて、7月、WHO 事務局長は、緊急対応枠組み (Emergency Response Framework: ERF) に基づきグレード3を宣言し、WHO の活動における優先順位を上げた。そして、8月、WHO とギニア・リベリア・シェラレオネは、合同対応計画を立ち上げた。この段階になって、ようやく、WHO 事務局長は国際社会へ向けて PHEIC を宣言した (WHO Ebola Interim Assessment Panel 2015)。

ギニアでは、その数日後、大統領が非常事態宣言をおこない、エボラ対策国家調整会議を設置した。エボラ対策調整官の指導力もあって、ギニア政府による活動がようやく本格化した。

WHO 主導による国際的対応だけでは不十分ということで、9月、国境なき医師団は「世界の指導者たちはエボラ出血熱への対応に失敗しつつある」との見解を緊急発表した (MSF 2015)。その後、国連事務総長の主導により、前例のない、UNMEER が設置された。

その後、各国の努力に加え、国際的対応が拡充された結果、ついに 2016 年 1月、WHO は、西アフリカにおけるエボラ出血熱の流行の終息を宣言した。

(2) なぜ WHO は PHEIC 宣言に遅れたか？

国際保健規則 (International Health Regulations: IHR) に基づき、国際的な公衆衛生上の脅威となり、国際的対応をとくに

必要とするものについて、WHO事務局長はPHEICを宣言できる。これまで、4回宣言されている。2009年の新型インフルエンザの世界的流行、2014年の野生型ポリオの世界的流行、2014年の西アフリカにおけるエボラ出血熱流行、2015年のジカ熱の世界的流行である。このうち、2014年の西アフリカにおけるエボラ出血熱流行に際して、PHEIC宣言が遅れたという批判がある（Garrett 2015; Gostin and Friedman 2014）。

ギニア国内においては、当初、ギニア政府と国境なき医師団との間で、見解が分かれた。ギニア保健省は、2014年3月にエボラ出血熱の流行を宣言したが、国際的な対応が必要とは認識していなかった。それに対して、国境なき医師団は、国際社会に対して、同月に「地理的な広がりは前例がない」、6月に「もはや制御できない」と警告を発した。このような見解の対立のなか、WHOは、ギニア政府の意向を尊重したと言える。しかし、その数カ月後、首都のコナクリにまで感染が広がったこともあり、結局のところ、WHOはPHEICを、ギニア大統領は非常事態を宣言せざるをえない状況に追い込まれた。

PHEIC宣言が遅れた理由として、第1に、ギニア国内の要因が挙げられる。西アフリカで最初のエボラ出血熱流行だったため、ギニア政府に十分な経験がなかったことがある。また、南東部の森林地帯で発生しても、首都コナクリにまでは広がらないだろうという楽観論もあった。また、PHEIC宣言による経済的な打撃を恐れたことから、情報開示に消極的となった。この点で、早期から国際的対応を国際社会へ求めていた国境なき医師団との関係には緊張関係が生じた。

第2の理由として、WHOの組織体制が挙げられる。アフリカにおいては本来、WHOのアフリカ地域事務所（Regional Office for Africa: AFRO）が率先して支援を展開することが期待されている。しかし、人的・資金的な能力の制約から、AFROは効果的な役割を果たせなかった。また、ジュネーブに本部がある政府間機関であるWHOは、組織内的には7月にグレード3を宣言したが、対外的には、加盟国であるギニアなどの政府の意向を配慮して、8月までPHEICを宣言しなかった。

3月から国際社会へ向けて警告を発していた国境なき医師団から見ると、この数カ月の遅れが多くの命を奪う結果を生んだと言える。

(3) なぜ国連事務総長はUNMEERを設置したのか？

2014年8月にWHOがPHEICを宣言し、国際的対応が呼びかけられたからといって、すでに大流行していたエボラ出血熱が終息する見通しはなかった。9月、国連事務総長は、安全保障理事会と総会の決議を経て、UNMEERを設置した。国連事務総長の主導による政治ミッションや平和ミッションはよく知られているが、UNMEERは前例ない感染症対策のための国連ミッションとして2015年7月まで活動した。

UNMEERが設置された理由として、第1に、WHOの指導力と調整力に限界が見られ、国連として内外に強い指導力を示す必要があった。本来WHOの任務であるはずの感染症対策を使命とする国連ミッションを事務総長が設置することは、異例である。そのことは、時限的ではあるが、国連の専門機関の一

つとして自律しているWHOの上部に組織を設置すると受け止められたため、WHOからの抵抗を受けた。本来は主従関係のない国連事務総長とWHO事務局長との関係にも緊張が生じた。

第2に、人道問題であるとして、国連人道問題調整事務所（United Nations Office for Coordination of Humanitarian Affairs: OCHA）が対応に乗り出し、国連中央緊急対応基金（Central Emergency Response Fund: CERF）を活用するという案もあったが、それは実現しなかった。自然災害や武力紛争による人道危機が増えていたなか、感染症を起因とする新しいタイプの人道問題に対応する経験も実施体制も資金もOCHAにはないと判断された。

第3に、一刻の猶予もないなか、すぐに使える活動資金を大規模に動かす必要があった。そのためには、国連ミッションとして安全保障理事会と総会に承認してもらうのが近道だった。

UNMEERの評価であるが、WHOの指導力と調整力の限界が見られ、OCHAによる対応も困難とされたなか、大規模な資金調達のために他に選択肢はなかったと見られている。しかし、その活動が効果的であったかどうかについて、ギニア国内では非常に厳しく評価されていた。

(4) グローバルヘルス・ガバナンスを改善するための提案

これまで、グローバルヘルス・ガバナンスの改善へ向けて、すでにいくつもの提案がなされている（Katsuma, et al. 2016; Kruk, et al. 2015; Moon, et al. 2015; WHO 2015; WHO Advisory Group 2015）。ここでは、それらを踏まえながらも、ギニアの

事例から得られる教訓をもとに、新興感染症の脅威にさらされた人びとを置き去りにしないよう、グローバルヘルス・ガバナンスを改善するため、いくつかの具体策を提案したい。

第1に、IHRで義務づけられている危機管理対応のうえで最低限必要な能力（コア能力）の獲得など、国内的に実施するための国際協力が必要とされている。ギニアでは首都のコナクリにしかラボ施設がなかったため、そもそもエボラ出血熱の実態が十分に把握されていなかった。この点で、日本から供与された簡易検査器 Genie III（Kurosaki, *et al.* 2016）は可動式で地方でも使えたため、ギニア政府から高く評価されていた。

第2に、PHEIC 宣言に関して、最終的に WHO 事務局長が判断するにしても、その判断の根拠となる健康リスク評価は独立した常設的組織において科学的におこなわれるべきであろう。

第3に、各国における国連カントリー・チーム（United Nations Country Team: UNCT）を基盤として、国連開発援助枠組み（United Nations Development Assistance Framework: UNDAF）やテーマ別グループ（thematic groups）など既存の調整枠組みを活用しながら、感染症の流行に対応できるようにしておくべきである。医療・保健分野において主導的な役割を果たす機関は WHO で変わりない。しかし、ギニアでのエボラ出血熱は、感染者に祈祷した呪術師や、死亡者に触れて別れを告げた親族や友人によっても広がったため、医学や公衆衛生学に限らず、社会学や文化人類学を含めた学際的な対応が必要だと言える。

第4に、感染症の流行を伴う複合的な人道危機に発展してしまった場合、将来的には、UNMEER のような国連ミッション

は設置せずに、OCHAが対応に乗り出せるよう準備しておくべきである。将来の保健人道危機に対応できるように、感染症対策の経験がある国連人道調整官の派遣や、調整のためのクラスター制度の活用など、OCHAの能力強化が望まれる。

おわりに

　本稿では、社会的に排除されやすい脆弱な人びととして、暴力や人権侵害の脅威から逃れようとする難民と、新興感染症の脅威にさらされたエボラ出血熱感染者に注目した。SDGsの目標3「すべての人に健康と福祉を」に取り組むうえで、「誰も置き去りにしない」ことが重要である。脅威そのものを減らす保護の政策が重要であると同時に、脅威にさらされた脆弱な人びとの強靭性を高めるためのエンパワーメントが不可欠である。難民や新興感染症の脅威にさらされた人びとのための技術革新と教育を普及させ、脆弱な人びと自身が情報をもち、みずからプロセスに参画できるようにすることが、包摂的な社会をつくることにつながる。

　政府が、国内的に対応できていないのに、国際的な支援を要請しない場合において、国際社会として保護する責任を果たすためのグローバル・ガバナンスのあり方が問われている。国家主権と内政不干渉の原則は尊重されるが、社会的に排除されやすい脆弱な人びとが新興感染症などの深刻な脅威にさらされているときには、IHRのような国際的合意に基づく迅速な国際的な対応が不可欠であろう。

謝辞と付記

本研究は JSPS 科研費 JP25560389, JP26245023 と、早稲田大学特定課題 2016K-330, 2017B-333 と、国際医療研究開発費 29 指 2003 の助成を受けたものです。

なお、本稿は、『アジア太平洋討究』の No.32 に掲載された日本語論文と、No.28 に掲載された英語論文をもとに、加筆・修正したのち、日本語で執筆したものです。

注

1 国連 UNHCR 難民映画祭（Refugee Film Festival）2017 については、以下のウェブサイトを参照。http://unhcr.refugeefilm.org/2017/
2 『はじめてのおもてなし』の日本での公開上映については、以下のウェブサイトを参照。http://www.cetera.co.jp/welcome/
3 『難民と移民に関するニューヨーク宣言』の日本語訳は、以下の国際連合広報センターのウェブサイトよりダウンロードできる。http://www.unic.or.jp/files/a_71_l1.pdf
4 「BBC 世界の諸問題と子どもたち〜貧困・紛争・暴力にさらされる子どもの権利を考える」（丸善出版、2017）は、第 1 巻『21 世紀の子どもたちの未来〜国連ミレニアム宣言から 15 年後』、第 2 巻「ガザの紛争下の子どもたち」、第 3 巻「ボコ・ハラムに拉致された少女たち」の合計 3 巻から構成される。詳しくは、以下のウェブサイトを参照。http://pub.maruzen.co.jp/videosoft/news/2017/BBC_worldchildren.html
5 UNRWA の活動を日本語で簡潔にまとめた文献として、清田・服部（2016）がある。

「誰も置き去りにしない」 125

6 UNRWAとその活動分野については、以下のウェブサイトを参照。https://www.unrwa.org/

7 JICAの平和構築分野における取組みについては、以下のウェブサイトを参照。https://www.jica.go.jp/activities/issues/peace/approach.html

8 ジャパン・プラットフォームとその仕組みについては、以下のウェブサイトを参照。http://www.japanplatform.org/

9 報道発表「ミャンマー・ラカイン州北部における情勢不安定化を受けたミャンマー及びバングラデシュに対する緊急無償資金協力」は以下のウェブサイトを参照。http://www.mofa.go.jp/mofaj/press/release/press4_005076.html

10 「『母子手帳』世界の動き〜第10回母子手帳国際会議に寄せて(2016年11月23-25日、東京)」による。以下のウェブサイトを参照。https://www.jica.go.jp/topics/feature/2016/161118.html

11 小泉総理訪問時の日本の対パレスチナ支援の一部として、日本政府からUNICEF経由の「小児感染症予防及び栄養状況改善並びに新生児の院内感染予防計画」(337万ドル) がパレスチナへ供与された。報道発表は以下のウェブサイトを参照。http://www.mofa.go.jp/mofaj/press/release/18/rls_0713b_2.html、英語：http://www.mofa.go.jp/region/middle_e/palestine/assist0607.html

12 Monthly JICA 2008年7月号「特集 母子保健〜かけがえのない命をまもるために」で紹介されている。以下のウェブサイトを参照。https://www.jica.go.jp/publication/monthly/0807/01.html。また、JICAの2016年度トピックスの一つ「【命と健康を守る国際協力】母子手帳知られざるストーリー〜日本で受け取り故郷パレスチナへ、難民の母子に希望」として以下のウェブサイトで紹

参考文献

1 BBC(2017)「第3巻 ボコ・ハラムに拉致された少女たち」『BBC世界の諸問題と子どもたち〜貧困・紛争・暴力にさらされる子どもの権利を考える』丸善出版。

2 JICA人間開発部（2012）「母子保健事業における母子手帳活用に関する研究〜知見・教訓・今後の課題」（人間 JR12-024）国際協力機構（JICA）人間開発部。

3 清田明宏・服部修（2016）「国連の難民救済事業〜 UNRWAの活動」臼杵陽・鈴木啓之編著『パレスチナを知るための60章』明石書店、pp.210-214。

4 Garrett, L. (2015). Ebola's lessons: How the WHO mishandled the crisis. *Foreign Affairs*, September/October Issue. [https://www.foreignaffairs.com/articles/west-africa/2015-08-18/ebola-s-lessons]

5 Gostin, L., and Friedman, E.A. (2014). Ebola: A crisis in global health leadership. *The Lancet*, Vol.384, No.9951, pp.1323–1325. [http://dx.doi.org/10.1016/S0140-6736(14)61791-8]

6 Katsuma, Y., Shiroyama, H., and Matsuo, M. (2016). Challenges in achieving the Sustainable Development Goal on good health and well-being: Global governance as an issue for the means of implementation. *Asia-Pacific Development Journal*, Vol.23, No.2, pp.105-125. [http://www.unescap.org/sites/default/files/chapter6_0.pdf]

7 Kruk, M.E., Myers, M., Varpilah, S.T., and Dahn, B.T. (2015). What is a resilient health system? Lessons from Ebola. *The Lancet*, Vol.385, No.9980, pp.1910-1912. [http://dx.doi.org/10.1016/S0140-6736(15)60755-3]

8 Kurosaki, Y., Magassouba, N., Oloniniyi, O.K., Cherif, M.S., Sakabe,

S., Takada, A., Hirayamam K., and Yasuda, J. (2016). Development and Evaluation of Reverse Transcription-Loop-Mediated Isothermal Amplification (RT-LAMP) Assay Coupled with a Portable Device for Rapid Diagnosis of Ebola Virus Disease in Guinea. *PLoS Neglected Tropical Diseases*, Vol.10, No.2. [http://dx.doi.org/10.1371/journal.pntd.0004472]

9 Moon, S., *et al*. (2015). Will Ebola change the game? Ten essential reforms before the next pandemic. The report of the Harvard-LSHTM Independent Panel on the Global Response to Ebola. *The Lancet*, Vol.386, No.10009, pp.2204-2221. [http://dx.doi.org/10.1016/S0140-6736(15)00946-0]

10 MSF (2015). Pushed to the limit and beyond a year into the largest ever Ebola outbreak. [https://www.doctorswithoutborders.org/sites/usa/files/msf143061.pdf]

11 REACH (2014) "Understanding Social Cohesion and Resilience in Jordanian Host Communities: Assessment Report," Amman: British Embassy.

12 UN High Commissioner for Human Rights (2015) "Report of the United Nations High Commissioner for Human Rights on violations and abuses committed by Boko Haram and the impact on human rights in the affected countries" (A/HRC/30/67), Geneva: UN Human Rights Council.

13 UNICEF (2016) "Children on the Move, Children Left Behind: Uprooted or Trapped by Boko Haram," Dakar: UNICEF West and Central Africa Regional Office.

14 UNRWA (2017) "The Annual Report of the Department of Health 2016," Amman: UNRWA.

15 WHO (2015). Follow up to the World Health Assembly decision on the Ebola virus disease outbreak and the Special Session of the Executive Board on Ebola: Roadmap for action. [http://apps.who.int/about/who_reform/emergency-capacities/WHO-outbreasks-emergencies-Roadmap.pdf?ua=1]

16 WHO Advisory Group (2015). Advisory Group on Reform of WHO's Work in Outbreaks and Emergencies. [http://www.who.int/about/who_reform/emergency-capacities/advisory-group/face-to-face-report-executive-summary.pdf?ua=1]

17 WHO Ebola Interim Assessment Panel (2015). Report of the Ebola Interim Assessment Panel. [http://www.who.int/csr/resources/publications/ebola/ebola-panel-report/en/]

第4部　わたしのSDGsへの取組み

UNU/jfUNU ジュニアフェローシンポジウム 2017
「持続可能な地球社会を目指して：私のSDGsへの取組み」に
参加したUNU修了生30名のエッセイをセッション別に掲載いたしました。

セッション1
Session 1

持続可能な社会とグローバル・ガバナンス

Sustainable Society and Global Governance

Name	Christopher Archboy Agoha
Affiliation	Political Affairs Officer, United Nations Mission in Liberia
Nationality	Federal Republic of Nigeia
Courses	UNU International Courses
Year	2004

Ineffective Health Systems in Liberia
Improving Health Sanitation Education Programs among Communities Affected by Ebola Epidemics

1 Problem Statement
Disease control in Liberia is greatly hampered by poor sanitation, weak, or in some cases non-existent health systems.

Liberia health sector was crippled during the civil war, with facilities destroyed, loss of skilled personnel and lack of medical supplies. Water, sanitation and hygiene (WASH) infrastructure suffered deep destruction and collapse. The sanitation situation is worse, with only 25 percent of households (53 percent urban and 17 percent rural) having access to improved sanitation (AfT, 2011). The prevalence of open defecation (77 percent of rural households and 30 percent of urban households) and lack of solid waste disposal or sewage systems, poor drainage and disposal of garbage all pose contamination threats to drinking water. Also, 43 percent of household in Liberia did not practice hand washing with soap, while 62 percent do not wash hands after using the toilet and a further 68 percent reported that they do not wash hands before eating (AfT, 2011). These trends put the health of communities at risk, especially children and the elderly. The Ebola crisis in Liberia not only created a public health emergency, but also exposed a series of underlying weaknesses and vulnerabilities within the country's health systems.

2 Causal Analysis
In the capital city Monrovia and its immediate environs, communities living in urban slums were heavily impacted by the Ebola crisis. The high density population has no knowledge about the Ebola epidemics and refused to adhere to prevention and treatment protocols.

Traditional funeral rituals of washing the dead were a risk factor in the spread of Ebola as the body is at its most contagious stage post-mortem. Policy to cremate all bodies of suspected Ebola victims in Monrovia was seriously resisted. The attitude towards Ebola was rather conspiratorial as a *Liberian Observer Newspaper* published that the Ebola virus is genetically modified organism to be used as bio-weapons on Africans in an attempt to reduce Africa's population. While some members of communities did not believe that the disease existed, they thought it was 'invented' by the government to collect money from international donors (Washington Post 2014). The central argument here is that, in spite of the human devastation (4,806 deaths) caused by the Ebola epidemic, there is still a narrow and linear way of thinking about improving and strengthening of the health systems in Liberia (Denney et.al 2015). Accesses to health facilities are severely limited, and as the Minister of Health remarked; "Our health infrastructures were not designed to cope with the kind of outbreak that we had".

3 Policy Action to improve and strengthen health systems in Liberia

The adoption of health sanitation education policy that targets communities and focuses on the following areas:

- Sanitation, hygiene, safe and appropriate health care
- Creation of high-functioning health workforce
- Utilization of strong health information system that provides reliable health data
- Access to quality medical supplies
- Support from a successful health financing system
- Innovative problem solving through the application of new technologies
- Elimination of harmful and dangerous traditional practices
- Effective leadership and governance

Reference

- **AfL – (2011) Agenda for Transformation: Steps Towards Liberia Rising 2030, Republic of Liberia**
- **Lisa Denney et.al (2015) After Ebola: Why and How Capacity Support to Sierra Leone's Health Sector Needs to Change, Overseas Development Institute (ODI) London**
- **Grace Fletcher (2015) The Aftermath of Ebola: Strengthening Health Systems in Liberia.**
- **The Washington Post (2014) "Major Liberian Newspaper Churning out Ebola Conspiracy and Conspiracy", Washington D.C.**

Name	Allan Volante Cledera
Affiliation	San Beda College, Manila, Philippines
Nationality	Republic of Philippines
Courses	UNU Global Seminar – Shonan Session / Director, UNU Alumni Association
Year	2006

"Global Change and Resilience"

Microfinance and Climate Smart Agriculture: Integrated Farming System and Social Business

Sustainable agriculture in the developing country like Philippines is confronted with more challenging issues to date. From the rural-agricultural development towards global food security issues, agriculture has been in the forefront of development agenda. However, elevating the agricultural-rural communities to viable and sustainable growth proves to be a very complex and hard struggle. The farmers and its communities still faced with the cyclical problem of poverty. Many of the farmers are always in troubled with the "income-gap" in each planting season, thereby, bringing them to a level of "perpetual-indebtedness" to the suppliers and traders of inputs and output. Aside from the profitability issues, the farmers are also directly confronted with the climate change problems, rendering them more vulnerable to poverty and putting them to the highest risk in the financial sector to invest in.

Technology in agriculture has brought production to its peak but weak financial access on capital and post harvest facilities to marketing of their produce, has always brought the farmers into a state of bankruptcy in each cropping cycle. The rural-sector, which is basically agriculture-based communities are slowly migrating to the urban communities to find "better livelihood". Despite the "best" effort of the government to support the agriculture sector, it continues to struggle to bring down its programs and services to the grass root level.

It is under this context that I have developed and pilot tested a program through a financial intermediation scheme, the "microfinance and climate smart agriculture: integrated farming system and social

business (IFS-SB)". This aims towards:
1 Full time farmers

Although the farmers work is full time, they face the income gap from planting to harvest (3-6 months). Adopting integrated farming system under a "ladderized" methodology, they are able to earn on a daily basis through multiple harvest from major crops, short- termed minor crops like vegetables and livestock, as well as other crops such as herbs and spices, fruits, mushroom productions etc. From a regular production loan per hectare of 1,000 USD the farmers are now able to access loan capital up to 5,000 USD per cropping cycle. Thereby, increasing their man-hours of 2-3 hours a day to 8-10 hours a day, also engaging 2 to 3 household members with proper income.

2 Price Stability

More often the farmers would complain about the high prices in input and low prices on their produce. In designing and promoting social business through a collective purchases and delivery of inputs as well as sales of their produce, the farmers are able to get a better price from the suppliers and the buyers of their commodities. Promoting entrepreneurship among the farmers has encourage its families to become more "demand driven and market oriented" farmers.

3 Food security

It has been ironic to realise that many of the farmers are malnourished if not starving. Adopting the IFS-SB has enabled them to produce enough food for themselves, for their immediate community and spiralling towards a better value chain and market access.

The program is not perfect, and the learning curve is steep, but with increasing success stories from farmers, it will finally spread to other stakeholders to support and promote the program.

Name	Paul Agbanyima Ehigie
Affiliation	National Administration Coordinator, Save the Children International, Abuja, Nigeria
Nationality	Federal Republic of Nigeria
Courses	UNU Intensive Core Courses
Year	2013

SDG Target 3.8: Achieving Universal Health Coverage in Nigeria by 2030

Universal Health Coverage in Nigeria: The Way Forward

1 Problem Statement

There is this general saying that 'Health is Wealth', but current deplorable health situation does not indicate this realization. For example, the health insurance coverage currently operational in Nigeria covers only people in the formal sector, which is less than 5 % of Nigeria population. With this, the Nigeria health sector is therefore faced with problems of limited financial access, poor risk protection from health expenditures for the majority of the population and out-of-pocket expenditure on health as high as 65 % . In a population where over 70 % are living below a dollar a day, high expenditure on health increases the vulnerability of an already poor populace to slip further into poverty. Universal Health Coverage (UCH) is a health system where everyone, every disease and all costs are covered, aimed at ensuring that nobody becomes poorer as a result of seeking health care.

2 Causal Analysis

Non-achievement of Universal Health Coverage in Nigeria Could be as a result of the following:

1. Poor governance and lack of political will by government
2. No clear road map to Universal Health Coverage (UCH) in Nigeria
3. Poor attitude of citizens to health
4. Illiteracy and ignorance among citizens leading to recognition of health needs
5. Poor budgetary allocation to health sector
6. Health facilities are either not well equipped or not well managed
7. Human resources related issues in health sector like inadequate

health worker, negative attitude towards works, demotivated worker as a result of poor treatment and conflict of interest among health workers
8. Religious/Cultural beliefs by some citizen that discourages health seeking
9. No programme for hard to reach communities in the country (e.g. Nomadic herdsmen)

3 Proposed Actions

Therefore, to achieve this target, there must be combined efforts from all stakeholders including individual households, business sector, CSOs, INGOs, government and international corporations where everyone understands his/her role and by doing the following:

1. Governance should be improved, with strong political will that ensures Universal Health Coverage
2. Government should develop a clear road map towards achieving Universal Health Coverage
3. Government should encourage Public Private Partnership towards attaining Universal Health Coverage
4. There should be attitudinal change towards health seeking and utilization through education
5. Only qualified health worker should be recruited into health sectors
6. Government should improve her health sector financing
7. Health facilities should be better equipped than it is currently
8. There should be re-orientation of health workers to improve their attitude towards their works
9. Health practitioners should be motivated through continuous trainings, better work environment and better salaries
10. Increase engagement with all stakeholders, especially women during the formulation of Health policies
11. Health education programs and facilities should be introduced to target hard to reach population
12. Government should support the establishment of social insurance policy and scheme

In summary, the move towards UHC should therefore be hinged on 5 strategic pillars which are: strengthened governance and coordination, adequate revenue mobilization, efficient fund pooling and financial risk protection, Citizens' health education and effective strategic purchasing.

Name	Masahiro Ito
Affiliation	MA, Graduate School of Asia-Pacific Studies, Waseda University (1st year)
Nationality	Japan
Courses	UNU Global Seminar – Shonan Session
Year	2016

Overcoming Three-Delays Factors to the Sustainable Society
In Case of Nepal Orthopaedic Hospital

'Three-delay-model' is a key factor to protect people's rights to health. This essay will first describe the three-delay-model. This model is important because it 'points to an approach which prioritizes practical, measurable interventions designed to improve the availability and accessibility of services, which should in turn mitigate factors which impede the decision to seek these services' (Sereen T. & Deborah M. 1994, p1092); second, this essay will introduce a successful model hospital in Nepal Orthopaedic Hospital (NOH) overcoming some factors of the delays; and finally, and what NOH model can do to build a more sustainable society.

'Three-delay-model' consist of three phase of delays: '(i) the decision to access care, (ii) the identification of – and transport to – a medical facility, and (iii) the receipt of adequate and appropriate treatment.' (Emilie J. C., Alexander P. S., Andrea G. T. & Lee A. W. 2015, p418) These three phases have three factors of barriers. '[T]he barriers… are distance, cost, quality of care and sociocultural factors.' (Sereen T. & Deborah M. 1994, pp.1093-1108)

Following academic discussion above, NOH can overcome several key factors of the delays: distance, cost, and quality of care. Not only can it solve such problems, it is also sustainable thanks to the monetary system. These examples show that NOH can be a successful model to provide medical services for free.

Firstly NOH adopts a fair medical fee overcome cost issue. NOH adopts fair medical fee. The rich must pay more than the general medical fees to subsidise the poor medical fees so that the poor can receive the same treatment free or less fees. It is similar to different bed fees or relation between low-cost-careers and full-service-careers. The richer pay for more luxurious services, but the core services like receiving treatment or transportation have no difference.

Secondly NOH solves distance problem. It has main building in Kathmandu, and conducts outreach health camps 'from the outset to help poor patients who are not able to come to Kathmandu for their treatment.' (NOH Homepage)

Thirdly NOH has good quality of care. The hospital has high-tech equipment such as c-arm fluoroscopes, computerized radiography, ICU ventilators and so on. It provides trauma surgery, deformity correction, joint replacements, spinal surgery, arthroscopy, physiotherapy and so on. It has 200 beds and enough workers in 2014 (Saju P.).

As conclusion, with those practices NOH able to help to contribute for sustainable society. It has limitations such as sociocultural factors, modernisation of infra-structure and so on. However, NOH shows a good example for three-delay-model, and transform basic hospital systems.

Reference
- Emilie J. C., Alexander P. S., Andrea G. T. & Lee A. W. (2015) Applying the lessons of maternal mortality reduction to global emergency health. Bull World Health Organ No.2015-93, 417–423. doi: http://dx.doi.org/10.2471/BLT.14.146571
- Saju P. Nepal Orthopaedic Hospital document (2015).
- Sereen T. & Deborah M. (1994) Too Far to Walk: Maternal Mortality in Context. Sm. Sci. Med. Vol. 38. No. 8, pp. 1091-1110: doi: http://dx.doi.org/10.1016/0277-9536(94)90226-7
- NOH Homepage. Out-reach Health Camp http://www.noh.org.np/index.php?ref=content&Cid=21

Name	Joseph Muiruri Karanja
Affiliation	PhD Student, University of Tsukuba (2nd year)
Nationality	Republic of Kenya
Courses	MSc, UNU-IAS Postgraduate Degree Programme / jfScholarship Recipient / Director, UNU Alumni Association
Year	2013

Pastoralism, Politics and Poaching: the PPP Nexus in Kenya
Wildlife at the Centre of Politics

Kenya has been in the forefront in the wildlife conservation and protection. Approximately 6% of her landmass has been demarcated as protected areas. There are over 140 private and community conservancies covering roughly 3 million hectares of land (Kenya Wildlife Conservancies Association, 2017). Despite these strict protective measures, curbing wildlife crime has been a huge challenge.

After experiencing a dramatic increase in poaching, Kenya has intensified measures to curb this crime. The 2014 wildlife act increased maximum penalty for wildlife crime from US $400 to US $200,000 (Wildlife Act, 2013). The wildlife crime prosecution was established under the auspices of the public prosecutor, and it has increased the number of convicted and prosecuted suspects. In 2015, Kenya also launched a wildlife forensic laboratory to aid poaching prosecution (WWF, 2015). In 2016, Kenya burned 105 tons of ivory and 1.35 tons of rhino horns, the largest amount of confiscated items in history (AWF, 2016). Nevertheless, challenges remain.

The current improvement owes largely to the cooperation between local authorities and pastoralists. In fact, cooperation from pastoralists is essential to curb wildlife crime as about 84% of Kenya's landmass is classified as arid and semi-arid lands, providing home to about 10% of the country's population, and 75% of wildlife. Most residents in these lands are nomadic pastoralists (accounts for 60% of the livestock), and some them have become poachers (Orindi et al., 2007; Oxfam, 2008).

Social unrest among pastoralists is somewhat attributable to the increase of poaching activities in remote Kenya. Cattle rustling has intensified conflicts among some pastoralists and led to the proliferation of illegal weapons (Kaimba et al., 2011). This has made it easier for poachers to obtain sophisticated illegal firearms. Their purposes have been to sell not only ivory and rhino horns but also bush meat.

During the dry seasons, armed pastoralists forcibly graze their livestock into private ranches, national reserves and private conservancies. In January 2017, for example, 10,000 armed nomadic herders with approximately 135,000 head of cattle trespassed conservancies and private lands in Laikipia County (The Guardian, 2017). In the process, they killed wildlife supposedly to "protect" their livestock. The majority of politicians from arid and semi-arid areas has supported this type of illegal grazing, making it harder to curb poaching activities.

In remote Kenya, there is a general perception that wildlife protection is for the benefits of foreign tourists. Many local people do not find much incentive to be involved in wildlife conservation initiatives. Considering this, more efforts are needed to show benefits for local people to protect wildlife. The disarmament of pastoralists is another challenge ahead as there is a lack of systematic efforts to achieve specific results in the near future. At this time, many are afraid that the disarmament of one community will empower armed communities and create power imbalance.

(Joseph Karanja, Kenichi Matsui)

Name	Milan Karki
Affiliation	Programme Officer, Technical Support to Karnali Employment Programme
Nationality	Federal Democratic Republic of Nepal
Courses	UNU Global Seminar - Shonan Session, UNU Global Seminar - Kanazawa Session
Year	2006

Ending Poverty in Western Nepal by 2030
Promotion of Social Protection through Public Works Programme

Poverty reduction has been a central focus of the national development plan since the Ninth Plan. It has been further emphasized along with employment generation from the Tenth Plan. The Poverty Reduction Strategy Paper (PRSP) and the Tenth Plan (2002-2007) embodied employment generation, in general, and targeted programs, in particular, as one of the four pillars of the PRSP. With the end of the Maoist conflict and the realization of the urgent need for social protection to the poor and vulnerable, the Government of Nepal announced the Karnali Employment Program (KEP) through the budget speech of 2006 with an initial sum of NRs. 180 million. KEP was initiated as a scheme with '*Ek ghar ek rojgar*' (one family, one employment) as its objective and provide 100 days of guaranteed wage employment to every household.

The aim of KEP is to reach out to poor households that do not have any employment opportunities or sources of income. The objectives of KEP are:

A) supplementing the livelihoods of the poorest and most vulnerable households in the Karnali region through employment, and

B) creating local public assets which contribute to enhancing local livelihoods in the longer term.

However, the KEP programme management and technical support capacity at the district level were inadequately resourced and weak. Further, the institutional arrangements for (a) project selection and implementation; (b) targeting the poorest and most vulnerable, and (c) the disbursement of KEP funds for paying the participants were not suitable for meeting the programme objectives.

Since 2013, Karnali Employment Programme Technical Assistance (KEPTA) funded by UK Department for International Development has been supporting the Government of Nepal (GoN) in developing and testing new approaches to employment-led poverty reduction. The technical support team developed and tested a revised model for the effective delivery of the KEP and implemented this in few Projects in the 2014-2015. As a consequence of positive results, GON decided to initiate, the process of rolling out the reform of KEP based on lessons learnt. The key elements of the revised model include:

(a) a larger number of days of work for poorer households instead of the former smaller payments for a larger number of households without targeting;
(b) pro-poor targeting;
(c) improved transparency through job cards, MIS, PFM (public financial management) system
(d) regular payments
(e) effective project management and technical supervision.

It was also recognised that effective introduction of the improved approach would require:

(a) changes in planning at the VDC, district and central levels;
(b) development of management, engineering and social mobilisation capacity, and
(c) establishment of institutional structures and practices at the central, district, VDC and sub-VDC levels.

For all the above reforms tested for effective delivery of the programme the Government has received a UK Financial Assistance of 6 Million Pound on wages and the GoN has also allocation for the same. With the increased budget on wages and no guarantee for the costs and support on reform the social protection programme may fall apart in ending poverty in Nepal.

	Name	Neema Robert Kinabo
	Affiliation	Assistant Lecturer, College of African Wildlife Management, Mweka
	Nationality	United Republic of Tanzania
	Courses	MSc, UNU-IAS Postgraduate Degree Programme (Yokohama)
	Year	2013

Take Urgent Action to Halt Biodiversity Loss
Understanding the Role of Urban Biodiversity

Urban areas influence biodiversity outside them as city dwellers need materials, food and water outside city boundaries. Recognition of the critical role played by cities in finding solutions for biodiversity loss is fundamental as they consume 75% of the resources; they cover 2% of the world's land and 80% of greenhouse gases emissions (GHG's) comes from cities (CBD,2007). This essay highlights the challenge of understanding biodiversity in urban environment and the important role of urban gardens and nurseries in enhancement of livelihoods and biodiversity around cities in developing countries.

Urban areas are connected to a larger ecosystem and don't exist in isolation. Urban dwellers interpret biodiversity as something related to rare species or a cultural concept influenced by experience, emotions, education, and cultural background. A common notion is that biodiversity is something you have outside the city boundaries, whereas green open space, parks and gardens are found within. In most developing countries particularly in Africa this is also reflected in planning as in rural areas biodiversity is treated as an important factor, whereas in urban areas is treated as an aspect of conserved nature fragments or green spaces. This can be a natural side effect of the focus of conservation on endangered species in their natural habitats by CBD.

High urbanization is expected to occur in developing countries with approximately 80% of urban population living in cities by 2030 (UNFPA, 2007). As rapid urbanization pose threat to biodiversity, the role of green spaces and gardens becomes increasingly important. The green spaces and gardens in urban environment provides habitat for variety of species and play important ecological roles. The complexity of urban gardens determines the abundance and diversity of vertebrates and invertebrates and the socio-ecological system that arises forms part of what is known as "ecology-of-cities" (Alberti, 2005, Daniels and Kirkpatrick 2006).

For a developing country like Tanzania Small and Medium Enterprises (SME) contributes one third of Gross Domestic Product (GDP) with about 1.7 million business related activities that employs 3 million people (Mbura, 2007). A nursery (comprising different species of plants trees or flowers) is a micro agribusiness social enterprise that contributes towards sustainable livelihoods in many urban areas in developing countries. Nurseries propagate and sell plants and trees for urban gardens, architectural landscaping, fruit and vegetable seedlings to meet local commercial and home garden demand. Forestry needs, either for individual woodlots or for community forests, are also often supplied by nurseries.

The income generated from nurseries can contribute to household access to food, payment for basic health services, and act as cushion to prevent the household from falling in extreme poverty condition. For women in developing countries nursery agribusiness provides a unique opportunity for them to engage in business and play a more active role in contribution of household income while providing greater flexibility (compared with employment) to combine work and household responsibilities.

The importance of urban biodiversity should not be underestimated as it improves quality of life in cities and represents patches of larger ecosystem.

	Name	Koffi Isidore Kouadio
	Affiliation	Regional Certification Officer for Poliomyelitis Eradication, World Health Organization, Regional Office for Africa (WHO/AFRO)
	Nationality	Republic of Côte d'Ivoire
	Courses	UNU Global Seminar – Shonan Session / Director, UNU Alumni Association
	Year	2006

Eradicating Poliomyelitis by 2020
Update on Polio Eradication Initiative in the WHO African Region

Polio Outbreaks represent a major challenge in achieving the Sustainable Development Goals (SDGs), specifically the health related goal number 3. The Global Polio Eradication Initiative (GPEI) is a public-private partnership led by national governments with five partners – World Health Organization (WHO), Rotary International, the US centers for Disease Control and Prevention (CDC), the United Nations Children's Fund (UNICEF) and the Bill & Melinda Gates Foundation (BMGF). Its goal is to eradicate polio worldwide. The incidence of polio has decreased from 350 000 cases in 1988 in 125 countries to 37 cases in 2016 in 3 countries (Nigeria, Pakistan and Afghanistan). The date on onset of the last case of polio in the African region was on 21 August 2016, in Nigeria. In May 2012, the World Health Assembly declared polio eradication as a global public health emergency. The Polio Eradication and Endgame Strategic Plan 2013-2018 comprises of four objectives: (i) Poliovirus detection and interruption; (ii) Routine Immunization strengthening and Oral Polio Vaccine (OPV) Withdrawal; (iii) Containment and certification; (iv) Polio transition and legacy planning. With regard to my experience, I worked as a field surveillance officer for the "Stop Transmission of polio-STOP" program at WHO country office of the Republic of Congo. During my assignment, I contributed to the good health and well-being of the children by supporting the country in (i) strengthening active surveillance for acute flaccid paralysis (AFP) detection; (ii) improving vaccine coverage through routine immunization activities and several polio campaigns. I joined the

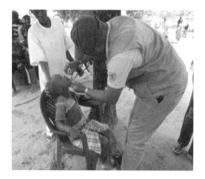

World Health Organization, regional office for Africa (WHO/AFRO) in 2014 as Polio Certification Officer and secretariat of the African Regional Certification Commission (ARCC) for poliomyelitis eradication. I am assisting the commission in reviewing country certification reports and complete documentations claiming a polio free status. I am supporting country's programs through: (i) capacity building of their national polio committees; (ii) technical and verification visits to ensure country complete documentations meet the required standards of the ARCC; (iii) and advisory services.

Based on existing criteria, 38 out of 47 countries have their complete documentation accepted by the ARCC, the only body to oversee the certification process and to certify the region free from polio. Despite all efforts, the region is facing many challenges: (i) Lake Chad basin polio transmission; (ii) Surveillance gaps at subnational level; (iii) Low population immunity; (iv) Insecurity and inaccessibility; (v) Population movements; (vi) Outbreaks of vaccine derived polio virus; (vii) other emerging outbreaks (Ebola, Yellow fever); (viii) Financial gaps to complete and maintain polio eradication.

The key priorities in polio eradication activities towards achieving SDGs in the region are: (i) to stop the transmission of polio in Nigeria, improve AFP surveillance and quality of campaigns in poor performing areas; (ii) to strengthen routine immunization program; (iii) to ensure timely phased containment of polioviruses and support countries documentation of polio-free status; (iv) to ensure proper documentation of polio legacy and best practices for strengthening other public health programs.

Name	Yuichi Kubota
Affiliation	Lecturer, University of Niigata Prefecture
Nationality	Japan
Courses	UNU International Courses
Year	2006

Promoting Peaceful and Inclusive Society
Public Service Provision and Welfare to Generate Integrated Identity in Post-conflict Sri Lanka

One of the challenges for sustainable society is the reconstruction in communities that experienced armed civil conflict. Post-conflict reconstruction is the highest priority on the agenda for political leaders, international donors, and locals in those communities.

Development programs tend to emphasize the physical construction of housing and infrastructure because the conflict causes enormous damage to human life and its socioeconomic foundations. However, the psychosocial dimension in post-conflict reconstruction is as important as conventional humanitarian assistance. Problems that make post-conflict reconstruction onerous include the diversification of identity of civilians. The aftermath of ethnic conflict, for instance, is marked by an identity split along ethnic lines. The disunion of subnational groups and communities is likely to occur due to a trust deficit in which people trust their co-ethnics more than out-group members.

My recent research sheds light on such influence of armed conflict over civilians and explores why and how individuals' wartime experience continues to influence their identity in post-conflict society. Conflict parties interfere in civic life in various ways. An example of this is the establishment of governance by anti-governmental armed groups under which the groups autonomously manage the relation with civilians. Activities of such governance include not only the use of coercion but also the provision of public services to civilians such as law-enforcement, security, education, health, welfare, economic well-being, and banking.

Sri Lanka is a case where the Liberation Tigers of Tamil Eelam

Figure 1. Wartime Experience of LTTE Governance and Post-conflict Subnational Identity

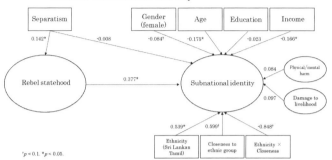

Source: Kubota, Yuichi. 2017. "Imagined Statehood: Wartime Rebel Governance and Post-war Subnational Identity in Sri Lanka." World Development 90, p.208.

(LTTE) developed institutions of governance in northern and eastern parts of the country. Findings of my case study suggest that civilians' memory of LTTE service provision is linked to the formation of a subnational identity in the aftermath of conflict (Figure 1).

The legacy of wartime experience persists and retains an impact on civilian identity in the post-conflict context. I propose that those charged with the task of post-conflict reconstruction need to take into account the long-lasting influence of civilians' unusual experience in order to successfully rebuild an integrated society. The post-conflict government should promote nation's welfare through an adequate supply of public services for those who were affected by the wartime politics. The notion that the northeast was underserviced by the central government encouraged a view supportive of rebel governance. Civilians may have felt that they would receive more services from the rebels. Those civilians should expect the government to provide better services than those they were unable to receive because of the conflict. The state will not be able to generate a national identity if it dismisses the influence of rebel statehood upon which civilian identity was based.

Name	Guilherme Cruz de Mendonça
Affiliation	Rio de Janeiro Federal Institute (IFRJ)
Nationality	Federative Republic of Brazil
Courses	Fellowship UNU-IAS(Yokohama)/ Visiting Doctoral Fellowship UNU-ISP
Year	2012

Education for Sustainability: BioCultural Diversity in Cities, Governance and Knowledge.

Nature - society relationships, expressed in the concept of 'BioCultural Diversity' (BCD), are key factors to understand sustainability. BCD "comprises the diversity of life in all of its manifestations – biological, cultural and linguistic – which are interrelated (and likely co - evolved) within a complex socio - ecological system" (Maffi and Woodley, 2010). Ecosystem, species and genetic richness are interrelated to traditional knowledge, cultural values, practices, institutions and languages. These connections are BCD dimensions, which are important to maintain life. Literature indicates that rupture between nature and culture in Western thought is one of the environmental crisis cause. In addition, unsustainable patterns of development put BCD in risk and danger conditions. Half of world's population leave in cities and the impacts of urbanization process are of significant magnitude in cities limits and beyond. These are problems to achieve SDG 11 target 11.3 and SDG 15 target 15.9.

In the Western thought, modernity is characterized by a fragmented worldview. Nature and culture, subject and object are some examples. Life is complex and cannot be reduced to fragmented view of the world. It needs a holistic and systematic thinking that overcome the bridges of modernity. To articulate biological and cultural diversity in cities falls into the strategies to face the crisis. Considering its relevance to life, BCD should be at the heart of governance at all levels. The study describes international legal framework that established rights and obligations related to BCD. States Parties shall establish conservation policies and mainstream them into other public policies at national level. Constitutions of countries with highest rates of BCD were analyzed. At local level, the city of Rio de Janeiro, Brazil, was selected as a case study, because it contains significant BCD, which is under pressures from development and sports mega-events; contains an innovative master plan that brought three innovations: local policy for cultural landscape, the right to landscape; mainstream of policies for BCD into urban policy. The research revealed the alignment of local, national and international Law. However, legal and governance approach has limitations because it doesn't mean that all factors that endanger diversity are being properly addressed or the behavior of actors has changed.

Therefore, education for diversity is needed. As a teacher, I am giving my contribution working with students of Environment Management on project about urban agroecology in Rio de Janeiro. We have mapped community gardens and organic production, especially in poor areas. The idea is to understand how these spaces are lively laboratories for biocultural diversity, but also their governance. As stated in target SDG 4 "By 2030, ensure that all learners acquire the knowledge and skills needed to promote sustainable development, including, among others, through education for sustainable development and sustainable lifestyles, human rights, gender equality, promotion of a culture of peace and non - violence, global citizenship and appreciation of cultural diversity and of culture's contribution to sustainable development". The implementation of it is my and our challenge.

Name	Arsene Honore Gideon Nkama
Affiliation	Senior Lecturer of Economics, University of Yaounde II
Nationality	Republic of Cameroon
Courses	Fellowship UNU-IAS
Year	2000

By 2030, Reduce the Adverse Environmental Impact of Cities by Paying Attention to Waste Management in Public Space

A Multidimensional and Multi-stakeholder Education Policy to Promote Cleaner Environment in Yaounde

1 Problem Statement

In Yaounde, the capital city of Cameroon, people are always clean. Their houses are neat and tidy. However, what contrasts with this individual appearance is that people drop litter anywhere, anyhow, without paying attention to public space environment. For example, eating banana often results to a banana waste on the road. Drinking yoghurt usually results to a yoghurt waste in the street. Those travelling by car transform the car to a waste producer that generates waste as the car moves. In Yaounde, as in many Cameroonian localities, waste speaks all languages: not only French and English, our two official languages, but also all our 200 to 300 mother tongues. Our cultural habits in public space seem to be convivial with waste, contrasting with our private life. Such situation destroys the city environment and the municipality efforts to clean the city remain unsuccessful, leading to important financial resources loss. Consequently, our private spaces are clean while our public spaces are not.

2 Causal Analysis

Poor environmental education is a key explanation to this situation. Those transforming the city to a waste dump can be divided into two groups. The first consists of those who unconscientiously dirty the city and are not ready to repeat their action after an appeal. The second consists of those whose civic education level does not allow them to measure the public impact of their action, so they are ready to repeat their actions even after an appeal. Both groups need education.

3 Policy Aiming at Giving Yaounde a Cleaner Face

Giving Yaounde a cleaner face would be grounded in a multi-sector and multi-stakeholder educational policy. Such policy would consist in different interventions at different levels for all populations.

Achieving better management of waste in Yaounde would require the participation of all sectors and all populations: children, youth, women, men, workers, unemployed, local authorities, central government, the private sector (both employers and employees), academics, city and rural dwellers, and even development partners. Each group would be targeted by an appropriate education programme or play a given role. For example, curricula would be adjusted accordingly to allow school boys and girls acquire environmental friendly behaviours. The private sector would acquire waste management knowledge through conferences, symposiums, workshops and coaching. Central government would show strong commitment by explicitly adopting clear orientations and give clear signals to populations. Such policy would be geared at the head of State level. Local governments would search for innovative solution. Development partners would support decision making to adopt good practices and share knowledge and technologies. Financing the policy would require both domestic and foreign resources.

With regards domestic resources, the government would integrate such policy in State budget and put highest importance on domestic resources. Public-private partnership would be important. The financial system would be ready to support the policy.

Name	Lino Sciarra
Affiliation	United Nations Stabilization Mission in Haiti (MINUSTAH), the Department of Peace Keeping Operation (DPKO), United Nations,
Nationality	Republic of Italy
Courses	UNU International Courses
Year	2001

SDG 5 – Gender Equality and Empowerment of Women and Girls:

Capacity Building Project to Support Haitian Female Political Candidates through the Ministry of the Feminine Condition and the Rights of Women

1 Problem Statement

In Haiti, women's political representation at national level remains a challenge. Over the last two years, the political scene was dominated by central and local elections, the first punctuated by violence and which resulted in a provisional government until a democratically elected president and legislators came to power in 2017. During this period, women were unable to position themselves effectively despite the constitution requires a 30 per cent minimum quota of women representation at all levels. In the current legislature, only a Senator out of 30 (3.3%) and three Deputies out of 119 (2.5%) are women. The percentage of women legislators is not yet representative enough to support initiatives that would promote gender equality, and address the high sexual gender-based violence and related impunity, the limited women's access to basic education and health, and the lack of economic opportunities that affect most women.

2 Causal Analysis

Traditionally, women's representation in Haiti has been among the lowest in the western hemisphere. The electoral system designed by the constitution requires candidates to be elected by absolute majority in their constituencies. The electoral decree of 2015 required parties to field 30 per cent of women candidates, but the two-round system does not favor female candidates where broader acceptance of women's representation is yet to be established. The few credible female candidates reported during the last campaigns, which were at times marred by political violence, specific security concerns, and lack of expertise and funds to run successfully.

3 Proposing Action to be Supported

The Ministry of the Feminine Condition and the Rights of Women (MCFDF) is responsible for providing the normative and policy framework on gender equality in Haiti. Although underfunded, the MCFDF remains a key partner in mainstreaming gender issues in institutions, policies, and programs. Through MCFDF a long-term program can be legitimately funded to support future female candidates and address the identified constrains faced in the recent past through projects to be implemented at national and local level by reputable and impartial national women's organizations. For instance, as MINUSTAH is about to close in October 2017, it may be worthy and cost-efficient to support the activities of the Open Day Follow-up Committee that during past years has been working with MINUSTAH to ensure a common platform for women leaders from all ten departments and their participation to capacity-building initiatives in the capital.

Name	Ranaporn Tantiwechwuttikul
Affiliation	PhD student, the University of Tokyo (3rd Year)
Nationality	Kingdom of Thailand
Courses	UNU Intensive Core Courses
Year	2016

Accelerating Solar Photovoltaic Adoption
Policy-induced Technological Change

Even though energy always has a significant role in the living and civilization of humankind, we somehow take it for granted. For centuries, the global energy consumption has depended heavily on coal, oil and natural gas. Not to mention the soaring anthropogenic greenhouse gas emission and its linkage to climate change, these resources are exhaustible and cannot solely guarantee the growing energy demand. Therefore, the alternative energy resources are in urgent need and solar energy is one good candidate due to its abundance and predictability.

Amongst leading approaches to harness solar energy, the photovoltaics (PV) offers the dazzling one-step conversion technique from sunlight to electricity, which is widely regarded as the most convenient energy medium for modern society. Despite the impressive PV technology advancement, there is plenty of room for technology improvement. However, solely science and technology (S&T) are not sufficient to make a significant change towards sustainability, the roles of people, policy and society are equally important. Therefore, my research focuses on two issues: 1) To analyse key policy mechanism to secure PV innovation as well as to promote PV technology uptake, and 2) To explore the PV technological hurdles, especially the recent research on Perovskite Solar Cells (PSC). Theoretical framework is based on the innovation process, while methodological approach is framed by sectoral systems of innovation.

While it is known that solar photovoltaic (PV) technology has been developed since 1960s and series of technological breakthrough have been achieved, the market adoption is still very limited and concentrated particularly on the silicon-based PV. One leading challenge for PV technology adoption into the fossil fuel-based electricity system is primarily the cost issue. Hence, policy intervention plays an initial important role. Regardless of economic downturn, the tariff regression is foreseeable globally. My literature review and field observations in five countries: Germany is world's leader PV installation; China is low-cost leadership of PV production; Malaysia and Thailand show uncommon PV installation pattern over-dominated by utility-scale projects (>1MWp PV system), and India demonstrates novel PV initiatives and alternative financing options. Despite the geopolitical uniqueness, the trend towards distributed system and innovative investment mechanism will certainly catalyse technological diffusion. Yet, the challenges remain not only to compete with well-established conventional technologies (particularly coal-fired power plant in developing countries), but also how to implement PV different technologies in unique settings in order to optimise its potential.

As SDG Goal 7 addresses on 'the affordable and clean energy', renewable energy technology adoption and further technology diffusion are needed to secure the growing energy demand. Inevitably, policy intervention plays crucial role, and the UN system formulation and operation address this key role of global governance. Furthermore, the Green Climate Fund is amongst green financing initiatives which require strong stakeholder engagements ranging from government entities to project owners and communities. Due to the nature of multidisciplinary research, I aim to understand the key policy mechanism involving PV industry and sincerely hope to contribute to policy implications where technological exploitation and exploration should be balanced for the sustainable futures.

セッション2
Session 2

地球環境の変化とレジリエンス
Global Change and Resilience

Name	MD Azim
Affiliation	Founder with The Poli
Nationality	People's Republic of Bangladesh
Courses	UNU International Courses
Year	2010

Responsible Production for a Sustainable Future

Combining Sustainable Material Use and Non-toxic Manufacturing Practices to Improve Environmental and Human Health in a New Start-up "The Poli"

The pressing environmental issues such as waste management, water pollution, air pollution and so forth are widespread, particularly in developing countries. Our consumption, from day-to-day basis, carries a massive footprint and exposes populations to critical health hazards.

Annually as much as 2 billion barrels of oil goes to the plastic bag industry which creates a massive carbon footprint and critical health hazards to both people and natural ecosystems. Then it takes from 400 to 1000 years for plastic bags to degrade (Bell & Cave, 2011). No doubt we need to be critical about the source and processes involved in mass consumption. Paper bags are not a sustainable alternative either. Its production requires 4 times more energy compared to plastic bags and generates 70 times more air pollution and 50 times more water pollution compared to plastic bags, and requires cutting down of 17 trees per ton of paper (Bell & Cave, 2011).

Bangladesh's leather industry is worth a billion dollars a year, but that value comes at significant environmental and human costs. The process of tanning leather hides, in most cases, is highly toxic. Chrome tanning is the most controversial yet widely used practice in the fashion industry. The chromium salts that are used in the chrome tanning are carcinogenic, persistent, and harms both the environment and human health.

Taking into consideration the above environmental issues, the Start-Up "The Poli" is offering the possibility to reduce the environmental footprint through careful selections of the products for the global market starting with the bag industry. One of the sustainable development goals of the United Nations is to ensure sustainable production and consumption patterns in order to protect the earth (SDG 12). The goal of this initiative is 'doing more and better with less', increasing net welfare gains from economic activities by reducing resource use, degradation and pollution along the whole life cycle of a product, while increasing quality of life (UN, 2015). The Poli shares the very same idea and responsibility towards the environment and society.

Firstly, the Start-Up is offering bags made from jute and jutton as a sustainable and environmentally friendly option to be used on a day-to-day basis which can range from shopping bags to laptop carriers. Jutton is a fibre blended from jute and cotton, i.e. of vegetative source, and thereby completely biodegradable in nature. It consists of mostly jute, which is the most sustainable natural fibre in-terms of energy requirements for its production and it requires very little to no fertilizers or pesticides. Secondly, the company is offering leather bags tanned with vegetable dyes. Planned products include backpacks and tote bags that attract fashion conscious people who also care for the environment. The vegetable dyed leather employs a mostly naturally processed method that does not contain harmful chemicals that harm the environment and human health. The Start-Up plan to build the brand through an inspiring blog, that will feature how the company is addressing social and environmental concerns.

Reference
- Acknowledgement: Kirsi Keskitalo (Finland), Olof Engström (Sweden) and Rachel Covington (USA), for their continuous support to develop the project.

Name	Isaac Botchwey
Affiliation	University of Ghana
Nationality	Republic of Ghana
Courses	UNU Intensive Core Courses
Year	2015

Quality Education
Teachers, Agents for Bridging Disparities between Rural and Urban Education in Ghana

1 Problem Statement

Disparities in rural and urban education in Ghana in terms of enrolment, infrastructure, tools and equipment, enabling atmosphere, and professional teachers has often resulted in the migration of teachers from the rural areas to the urban centers.
Low enrolment, dropouts, poor performance and cycle of poverty are the outcomes of such gaps between urban and rural education. Enrolment of students in rural areas compared to urban centers is usually low coupled with high frequency rate of dropout. Again, the agrarian nature of rural Ghana demands labour of which family members are used including children of school going age.
Further, children of poor families usually tend to be vulnerable; unexposed, and in many cases may not break the cycle of poverty unlike their urban counterpart.

2 Causal Analysis

Education is perceived to be for the rich in affluent in most rural communities hence the high turnout of drop out as students at both the basic and high school levels.
Attitudes of some teachers and unattractiveness of rural Ghana continues to widen rural and urban gaps in education. Teachers often prefer urban centers with all its amenities than rural communities with little or none. Moreover, the absence of these facilities makes rural schools unattractive in Ghana.
Problem of accessibility and utilization of education structures and materials in rural communities of Ghana pose a threat to bridging

the disparity between against urban schools. Schools in most cases are sited several kilometers from communities where pupils have to commute to and fro daily. Where schools are accessible, basic materials and qualified are not available.

Cycle of poverty in rural Ghana includes poor funding from government, civil society organization (CSOs). Schools in urban centers often receive financial assistance in different forms unlike that of rural schools.

3 Proposing Action to be Supported

Enhancing the policy to support, promote and retain qualified teachers in rural schools through compensation, travel allowances, and or financial incentives. Existing policies on rural education in Ghana should be enforced by government, districts, municipals, and civil society organization with at the teachers at the core. Further, the state must show keen interest in rural education.

Using education as a tool to break the cycle of poverty (Plessis, 2014). This would ensure knowledge and skills development for employment, income and livelihoods. Children from poor home can then make effective and efficient use of resources in their communities whereas enhancing their lives.

Through school based projects such as "*school on wheels*" which aims at improving rural education while reducing the gap between rural and urban education. Through, resourced voluntary teachers would move to less privileged rural schools quarterly to train, assist, mentoring and bring education to the very doorsteps of rural schools. This means education would be everyone's responsibility and an everyday activity.

Name	Olga Dziubaniuk
Affiliation	Teacher, PhD Candidate, Åbo Akademi University, Finland (4th year)
Nationality	Ukraine
Courses	UNU Intensive Core Courses
Year	2012

The Role of Private Sector for Achieving SDG Targets
Marketing and Promoting Sustainability Goals in Europe

From the marketing perspective, we look at sustainable development and challenges from three general perspectives: environmental and social spheres that are supposed to be interconnected or balanced with economic benefits for the businesses and society. Global changes of the last decades forced private sector to become proactive in response to the environmental and social challenges as well as they brought new business opportunities. Active involvement of the private sector (especially small-size business organizations) in the searching for solutions to the challenges results in rising awareness of sustainable consumption, response to the demand on sustainable technologies, and developing responsible and less harmful for environment business practices. However, in spite the fact that the business people are also a part of the society, sustainable and ethical business leading beyond the legal regulations is still not such a wide spread activity in both industrial and developing countries. People simply do not care! Although large multinationals conventionally are more involved into sustainable business development and promotion, the small companies only gaining their momentum. How to make businesses being more interested to contribute for the achieving goals of sustainable development and enhance their willingness to be involved in cooperation with non- and governmental organizations on this matter? How can companies sell the "problem" of sustainability to the consumers to boost the awareness of the sustainable development issues? Well, it is significant for companies to see the value to gain from sustainable business practices.

Philosophy of sustainable business has a number of benefits. This statement is supported by the authors' involvement in the research on Finnish and European companies engaged in the sustainable business practices and other international businesses that implemented corporate social responsibility. Naturally, adaptation to the global challenges increased the number of business organizations that have placed the sustainability at the core of their business model, which also increased the development of so-called social entrepreneurship organizations. Extensive research emphasizes that engagement in sustainable business can deliver a competitive advantage especially for the small-size businesses. Marketing of the sustainable goals achievement adapted to the realities of the developing markets is one of the major contributions that European business can make. Bringing businesses in to the areas of poverty boosts local economic development, which is also positively reflected on the social benefits and even human rights security. Modern startups can enjoy the opportunity to obtain financial investment if they develop sustainable solutions due to investment companies and "business angels" became more loyal to the responsible businesses.

In conclusion, a modern role of business organizations and their marketing communication is to create awareness on how to overcome global challenges. University education focused on sustainable values (especially in business schools) is strategically important to stimulate the future development of responsible business. Although academic research on sustainability in marketing and business administration fields became a common practice, there is a strong need for empirical research on the methods of achievement of the Sustainable Development Goals that are still on their way to success.

Name	Mujahid Hussain
Affiliation	Executive Director, PEDA International
Nationality	Islamic Republic of Pakistan
Courses	UNU International Courses
Year	2009

Combat Climate Change Impacts Magnifies Disaster Risk and Increase Communities Resilience by 2030

Improve Disaster Preparedness, and to Reduce Vulnerability to Disaster Impacts on Lives and Livelihoods, Particularly of Women and Other Vulnerable Populations

1 Problem Statement

Pakistan is prone to natural hazards and disasters, including floods, earthquakes, and drought. The GFDRR reports that 9 % of the national territory is at risk from two or more hazards, representing 40.1% of the population and 41.6 % of GDP.[1] According to World Bank data 2014, Pakistan is bearing around $7 billion of economic loss[2] annually which is 2-3 % of national GDP. Major meteorological disasters occurred in recent past years, indicating a pattern of increasing recurrence associated with climate change, 17 out of 40 districts lacks hydro-met data. According to recent DFID Scoping Study that Pakistan lacking localize forecast capacity (lower level at district and community level). Another major gap in early warning system is that climate information is not produced for different user groups; agriculture, water, fishermen, travelers, smallholder farmers, poor and vulnerable areas and communities etc. Extreme weather events are among the most harmful impacts of climate change which is an obstacle to majority of the population (at least 40%) to achieve sustainable development goals.[3]

2 Casual Analysis

The causes of meteorological and weather-related disasters are multifold such as global changes in climate and weather patterns, low level of preparedness at government and community level, no or poor link between development and community resilience, capacity gap in the lack of timely forecasting of risks and at large mitigation to disasters, increases the vulnerabilities of Pakistani people to hydro-meteorological hazards. The lack of preparedness capacity at local government and community level, the increase in the frequency and intensity of metrological/hydrological disasters compounded to the problem and a risk to push millions of people to poverty trap every disaster years and reap the progress towards the sustainable goal of building resilience communities and nations.

3 Proposed Actions

Sharing Weather information such as advisory, alerts, early warning is key elements of preparedness and managing of disaster risk and among the best options to mitigate the impacts and costs of such events. The project objective is to improve disaster preparedness and to reduce vulnerability to climate change and disaster impacts on lives and livelihoods, particularly of women, from extreme weather events and enhance resilience of vulnerable populations. (Goal 13 of SDGs). This project is piloting in 2 districts with population of approx. 2 million, both districts are high risk areas to meteorological hazards.

- Expansion of weather information system/platform to generate climate-related real time data and information/services to communities engaged in preparedness to save lives and safeguard livelihoods from extreme climate events
- Scaling up the use of modernized and sustainable AWS and strengthening communities' capacities for use of EWS/Climate information in preparedness and response to climate related disasters
- Build disaster preparedness capacity of local community members; farmers, women and other vulnerable groups
- Provide early warning lifesaving information to 2 million population during pilot phase in the selected districts of Pakistan

Reference
- [1]**GFDRR Natural Hotspots Study: A Global Risk Analysis 2005**
- [2]**Pakistan Disaster & Risk Profile,** http://www.preventionweb.net/countries/pak/data/
- [3]**Goal 13: Take urgent action to combat climate change and its impacts Climate change magnifies disaster risk and increases the cost of disasters.**

Name	Hidayat Ullah Khan
Affiliation	Chairperson/Assistant Professor, Department of Economics, Kohat University of Science & Technology (KUST), Kohat
Nationality	Islamic Republic of Pakistan
Courses	UNU Joint Graduate Courses
Year	2005

Improve Education, Awareness-Raising and Human and Institutional Capacity on Climate Change Mitigation, Adaptation, Impact Reduction and Early Warning

Promotion of Environmental Awareness, and Capacity Building among Women to Mitigate the Adverse Effects of Climate Change by Reduced Greenhouse Gases (GHG) Emission via Adoption of Fuel Efficient Stoves

UN Framework Convention on Climate Change Climate (UNFCCC) recognizes climate change (CC) as a formidable threat for the earth and its inhabitants. Greenhouse gases (GHGs), among others, are major contributors towards CC, with their current concentration and emission levels can increase global temperature by 1.5℃ in comparison to that of 1850 to 1900, raise 40-63 cm sea level by 2100, and cause calamities with global implications.

South Asia is among the most vulnerable regions to the CC related calamities. Particularly, Pakistan is included among the top-ten vulnerable countries to these calamities (Global Climate Risk Index, 2016) owing to its strong economic reliance on climate-sensitive sectors like agriculture and forestry. Particularly, the country's densely populated low lying areas, with almost half of its population, are prone to floods and other calamities.

In rural Pakistan, wood and biomass are used for cooking, and heating in rural Pakistan that cause environmental degradation, and adversely effects health. There is a need for adoption of clean and efficient ways capable of transforming energy use patterns, improving indoor air quality by reduced GHGs emissions, and lessening strain on the natural resources.

Reserved and/or protected forest, covering a meagre 4-5% of the country's area, is a principal source of fire-wood and timber. Save for billion tree project (BTP), conservation and/or afforestation efforts are almost non-existence in the country that are crucial curbing ill effects of GHGs and CC.

To achieve the afore-mentioned objectives, a pilot project has been implemented in district Haripur, Pakistan, an agro-based rain fed region. The district is confronted with deforestation, stemming from demand hike for fire-wood and timber in the absence of concerted efforts through use of community-based conservation and afforestation efforts. The local people are aware of importance of forestry and natural resources owing to their primordial dependence on the same.

A local non-governmental organization (NGO) called Pakistani Hoslamand Khawateen Network (Pakistan's Courageous Women Network, English translation; abbreviated as PHKN), women focused and women driven NGO. PHKN, took a bold step, under the auspices of UNDP-EF-SGP, to promote environmental awareness, and community-based conservation practices among children, women, and local communities in five Union Councils of Haripur during 2009-10.

Under this initiative, capacity of around 3,200 women regarding environmental awareness, sustainable use of forest resources, and fuel efficient stoves (FES) making was built. Moreover, capacity of 1000 school children regarding basic health and hygiene, and forest and environmental conservation has been built. Besides, under community-based plantation campaigns 37,850 trees were planted through.

A project focusing on the long-term impact of the project on fire wood consumption, indoor air quality, and health of the women in district Haripur is proposed, which is aligned the SDGs on "Climate Change" particularly its goal of "improving education, awareness-raising and human and institutional capacity on climate change mitigation, adaptation, impact reduction and early warning". Moreover, up scaling and/or replication of FES is aligned with goal 7 of SDGs regarding "Clean Affordable Energy". The proposed project is expected to directly contribute towards "Reduction of GHGs emissions".

Name	Niels Burkhard Schulz
Affiliation	Environmental Expert Consultant, United Nations Industrial Development Organization (UNIDO)
Nationality	Federal Republic of Germany
Courses	Fellowship UNU-IAS (Yokohama)
Year	2004-2005

Bamboo for Urban Infrastructure
Innovate and Appropriate Technologies to Address Impacts of Urbanization Dynamics

The sustainability transition, or fight to achieve the SDGs, will be won or lost in urban areas: Cities are the locations where global populations and resource demands are increasingly concentrated. They cover less than 2% of the terrestrial surface, but now accommodate over half of the human population. Consuming about 75% of the material- and energy resources, urban areas are responsible for an even larger share of greenhouse gas emissions. Global populations are still growing, and by 2050, when nearly ten billion people will live on this planet, more than two-third of them will be urban. City populations are to double between 2000 and 2050 and at current resource intensity per capita, this development pathway is unsustainable. New ways to build cities need to be found.

Cities are open systems and their functional reach and impact exceeds well beyond the city limits: large tracts of rural lands are necessary to provide food, feed, fibre and other bio- resources for the consumption of urban dwellers. Construction materials and industrial minerals are extracted mainly in rural areas for processing in urban manufacture. Also with their demand for aquatic resources and clean air, cities indirectly appropriate ecosystem services from all over the planet, including for the dissipation of waste streams. While the urgency for urban sustainability solutions was now recognized as SDG 11, their setup and function indirectly affects attainability of most other SDGs, in particular nr 9, 13, 12, 7, 6, 8.

Technological innovations can provide solutions. For example fast growing, highly renewable and carbon negative building materials like modern bamboo composites have the potential to substitute a good share of demand for energy and emission intensive construction material such as steel for future cities. Bamboo belongs to the plant family of sweet grasses, and grows wood like biomass with outstanding tensile fibre strength properties. It grows faster and is more water and nutrient efficient than most other plants and can be cultivated on sloping and degraded lands. Innovations in wood processing technology now allow the manufacture of long lasting products and timber substitutes from bamboo to reduce logging pressure on existing forests. It creates rural employment opportunities to ease urbanization pressure, without competing with land required for food production. Biomass waste from bamboo processing can serve as sustainable feedstock for biocharcoal production to sequester carbon and manage soil fertility. This aligns with the "Bonn challenge" of GPFLR launched in 2011 and the "4 per mille" initiative launched at UNFCCC COP 21 in Paris 2015. The geographic locations of current urbanization dynamics (in particular Africa and Asia) coincides well with the tropical bamboo belt, where large diameter timber bamboo can be cultivated. This also reduces the current import dependency for construction materials of most rapidly urbanizing countries.

There is need to update of national building codes to accept such innovative construction materials, and to showcase and demonstrate these technologies e.g. in examples of public infrastructure. UNU could assist to facilitate cooperation with Institutions like UN-Habitat UNEP, UNDP and UNIDO. A project like "bamboo for urban infrastructure" can easily be aligned with initiatives for climate change adaptation and mitigation, improvement of resource efficiency and circular economy, rural energy access, land restoration and reforestation, rural value chain development, agricultural diversification and watershed protection. UNU in particular could act as multiplier to share insights and lessons learned between countries.

Name	Natalia Estrada Sifuentes
Affiliation	Administrative Manager in a Peruvian Company
Nationality	Republic of Peru
Courses	UNU International Courses
Year	2009

Aquaculture of Trout in Titicaca Lake and Resilience Strategies

"Most of the impacts of climate variability occur in Mesoamerica and the Andes." FAO.

Peru is characterized by being a country with an extensive biodiversity, and, as most developing countries it depends highly on its natural resources. Due to increasing of population in the region, new technologies coming up, and more access to higher education, Puno habitants have seen aquaculture of trout in Titicaca lake as an opportunity to generate income from another activity besides agriculture. Many farmers saw an opportunity in the industry and shifted from agriculture to aquaculture, making it their main practice for economic income. Although inland aquaculture of trout is being done in all levels, from the artisan to the enterprise level, the study focused in artisan families and describes how climate change affects the activity and what strategies government and people are taking in order to diminish the effects of climate change. However, actors should not think only on the current effects of climate change but on its future effects.

Among the negative effects of climate change in Puno region we can find increasing of temperature in Titicaca Lake's water, droughts in farming lands, and "friaje" in the region affecting part of the population, particularly those living in the highest part of the region.

The studied productive unit presents a subsistence vulnerability. The increasing of in-land water temperature above the standard records produces the death of biomass that reduces every family production considerably. However, the effects of climate in the region are diverse. While in Titicaca lake the temperature of water increases, agriculture lands and highlands are affected by extreme frosts. This makes it very difficult to harvest reducing food and economic development from their second main economic activity. With the objective of overcoming these negative effects, families, regional institutions and government have created strategies as mechanisms of adaptation and mitigation.

Knowledge and training, government institutions, NGOs spread the risks about climate change negative effects. This knowledge is complemented with ancestral knowledge to reduce loss of production.

Planning, because the temperature varies for periods of time, farmers plan their production in such a way that they will produce minimum quantities during seasons where the highest temperature is recorded.

Financing, credits from private banks and national financial institutions are granted to the artisan producers for the improvement of their equipment and cages.

Associations, cooperation is found among the productive units, this is one of the most reliable ways to overcome negative effects of climate change.

Climate change caused by either nature or human activities has a great impact in aquaculture in Titicaca Lake. Resilience is achieved by putting in practice strategies mentioned above and also by constant innovation of new productions models taking into account strengths, opportunities, weakness and, threats, and with participation of all actors in all areas geographically, socially, and governmentally.

Name	Jean-François Vuillaume
Affiliation	Researcher, Ignitia, Ghana
Nationality	Republic of France
Courses	PhD, UNU-ISP Postgraduate Degree Programme
Year	2012

Research Description on Global Change and Resilience
Private Sector Start-up and Weather Services in Developing Country

I graduated from the United Nations University in 2015 on sustainability sciences with a specialty on weather forecast optimization for flood early warning system prevention. Then I focused on working in developing country to gain knowledge on the practical implementation of my study at the United Nations University. My background is from physics with social sciences knowledge. I liked to challenge my view with ground reality related to international cooperation and private sector operating in developing countries. I started working for a private weather company in West Africa in 2015 using my knowledge to make weather forecast better.

As can know many students coming from university to the private sector is often a drastic change in focus and practice which have pro and con. However, it is an important step to understand better the interest of different actors. I was confronted with working in the "expat community" and trying to interact and getting knowledge from the local. Spending time understanding local culture is a crucial process to share idea and habit in a new country. I faced several issues on cultural difference which the United Nation University prepared me the best. I was surprised by strong inequality and that, money can often be a question depending some community. I also remain myself that international company whatever their size has known how to not follow the local tax system (It is true in all parts of the world).

Start-up company as I worked with share strong value on

innovation and technology rupture on the market. But for the forecast, most of the tools are already available for free, outside of having a cluster computer that isn't such expensive now-day. It lets me with a mix feeling that most of West Africa is largely capable of making their own forecast. Therefore, such issues are not covered enough within both the local university and the international community. Therefore, a real cooperation model that can both improve local living and permit to build a stronger state while getting complementary services from the private sector have to be think further.

セッション 3
Session 3

グローバル・シティズンシップ
Partnership for Global Citizenship

	Name	Tijana Kaitović
	Affiliation	- Business Development and Growth Manager, Farmia
		- Community Manager, Network of Change
		- Founder, Youth Energy of Serbia
	Nationality	Republic of Serbia
	Courses	MSc, UNU-ISP Postgraduate Degree Programme / jfScholarship Recipient
	Year	2011

By 2030, Empower and Promote the Social, Economic and Political Inclusion of All; Ensure that All Learners Acquire the Knowledge and Skills Needed to Promote Sustainable Development; Combat Inequalities; and Enhance Global Partnerships and Mobilizes All Available Resources

Strength through Solidarity

Humanity in the late 20th and 21st century was torn by many conflicts, human sufferings, economic and political crisis, and massive violation of human rights. Torn nations are still feeling the aftermath and consequences of this time, especially in the areas of statelessness, women's rights, government accountability, corruption, lack of education and economic opportunities and high levels of unemployment. Despite the best efforts of the international community including the United Nations-UN to build sustainable peace and development in these countries, the state-centered approach has continually proven unsuccessful.

Empowering the individual is a core principle in the process of real-world problem solving. My experiences as a founder have taught me that the best way to grow peace is not through state-centered approaches, but through individuals building sustainable communities and local economies through entrepreneurial endeavors. Growing up and working in post-conflict Serbia heavily influenced my interest and passion in seeking ways of sustainable peace-building and community development. As the first woman in my family to attend university, I focused my studies on the UN peace-keeping and peace-building missions I grew up with. I worked with NGOs, institutes, and universities around the globe and deeply studied SCR 1325 and MDGs. However, I returned to my home country for a short time, and realized that while it was at peace, opportunity for youth and others in society had not recovered. Serbia remained behind with 44.2% youth unemployment and the problem of emigration and 'brain drain' where many educated young people were leaving the country (UNDP Report). State-centred approaches from new, weak governments failed to manage old and new challenges facing the country. I realized that human development was crucial, in particular strengthening the role of civil society and the individual.

As a values-based leader, I focus on creating social change by establishing global partnerships. My most impactful techniques leverage education, gender lens, human-centered design and social entrepreneurship as tools for igniting individuals and (re)building inclusive global communities. After enabling social change in communities in India, Japan, USA, the Philippines, and my home region in the Balkans, I realized the need for peer support in social change and with a team created a transnational network of social changemakers. We founded an international and inclusive network of spicy, grass-root, and diverse social initiators who want to transform the world, community by community. Among our goals is promoting self-sustainable economic growth through sharing examples of good practice and solutions to challenges, specifically focusing on marginalized communities. This platform succeeded in 'erasing' boundaries between people and it provides a network where everyone has the opportunity to build a life of value. This experience affirmed my belief that countries need to take proactive action in order to successfully reach the SDGs. What is important is no longer preserving the artificial boundaries between people, but rather protecting and empowering the individual and creating global collaborations as core principles in the process of addressing real-world problem solving.

The way forward in building stronger and sustainable communities is in building innovative, inclusive and people-centered economies. A profound realization and hands-on experience has led me to the interdisciplinary research-based approach which puts individual human being in the center and nurtures the culture of partnerships and collaborations.

Name	Fatemeh Keyhanlou
Affiliation	Assistant professor, Islamic Azad University
Nationality	Islamic Republic of Iran
Courses	UNU International Courses
Year	2003

International Sanctions *v.* Partnership for Global Citizenship (PGC)

While long lasting international sanctions may sometimes result in changing public policies, their negative unforeseen consequences, can evidently be traced in the daily lives and behaviors of the population of the targeted State.

Since the Islamic Revolution in 1979, Iran has persistently faced various international sanctions (general and targeted) imposed by Western States (US & EU) and United Nations Security Council. Some of these sanctions were lifted pursuant to the Joint Comprehensive Plan of Action (JCPOA) singed in 2015 and some are still in force due to the alleged violations of international law by the government. These sanctions have revealed indirect but widespread effects on the enjoyment of basic human rights by the Iranian population. Financial and banking sanctions and blockades on the export of oil, for instance, affected the Iranian citizens' right to life through proscription the renewal of aviation industry and the subsequent air crashes and loss of lives. Bankruptcy and overmuch fired workers, inflation and economic crisis, malnutrition, limited access to adequate health care and drugs for special deceases were some other imminent consequences of sanctions affecting vulnerable classes of citizens.

Under these circumstances, the notions of social commitments, humanism and national solidarity have gradually been weakened in the society. The survival struggle became a critical concern for even the middle class citizens and their reliance on government and its power to restore the national prosperity had significantly decreased. Social and political dissatisfactions had been intensified but despite the expected function of sanctions, *i.e.* changing the behavior of the government, and instead of fundamental reforms, more pressure was imposed on the critics. Meanwhile, evading sanctions became a rich source of income with the inner layers of sovereignty being converted into defenders of sanctions and their continuance.

The atmosphere in which Iranians were and still are considering themselves as the most vulnerable victims of international sanctions, and look with pessimism to the commitments created by Western States (even in the framework of International Organizations), any discussion or education on partnership for global citizenship seems to be too absurd and abstract to be comprehended. To change this perspective, the notion of GC should be promoted with care and prudence. The starting point is restoring confidence and solidarity to national and international relations.

On the realm of international relations, commitment to JCPOA is the main step towards breaking out of international isolation of the last four decades and its full performance by the parties can undoubtedly amount to the ruin of distrust in inter-State interactions.

On national level, educating, training and public awareness are among priorities. Strengthening the spirit of cooperation, common responsibility and confidence building are of importance. Scientific, artistic, cultural and sport exchange programs play a clear role in decreasing mutual pessimisms and increasing mutual understandings. Dialogue and interaction are among the required elements in the revival of humanitarianism among Iranians themselves and with foreigners and should be supported.

	Name	Kaoru Kojima
	Affiliation	Student Consultant, Educational Affairs Department, Osaka Seikei University and Osaka Seikei College
	Nationality	Japan
	Courses	UNU International Courses / Director, UNU Alumni Association
	Year	2000

Fostering Global Citizenship for 2030 Agenda
Human Resources Development for Global Citizenship in a Field of Education, Japan

In Japan, practical uses of English had been ignored for a long time. On that account, most Japanese people are weak at English. Especially they tend to avoid communicating with English speakers in English or languages other than Japanese. A Global Citizen who should communicate with each other for understanding and respecting each other's individuality in this Earth, regardless of race, nationality, gender or age differences. Language is a capital instrument in communication for humanity. To be a Global Citizen, you need to use one of official languages of United Nations proficiently. As you know, the official languages are Arabic, Chinese, English, French, Russian and Spanish. Japanese is not applicable to the official languages. The official languages are not mother tongue for Japanese people. Above all, the English language has now become the universal language. Using English is the fundamental basis for being a Global Citizen. It requires that students who can acquire practical English skills in Japanese education urgently.

Against this background, it is important to be clear about the possible cause of poor English practical communication skills in Japanese education. In compulsory education of Japan, English is a required subject. However, there are not enough teachers who can communicate with English speakers in English and guide their students for practical uses of English. That is reason the most students in Japan are not good at using practical English. The young students have infinitely possibilities of finding solutions to the many global problems facing humanity. It therefore requires that we educate and foster young people who can contribute their skills, experiences and knowledge to be a Global Citizen. For attaining the targets, it is necessary to use the practical English language which is a means of communication.

My focus is on what is actually going on in my familiar places. I work in many educational facilities; some private educational institutions which are composed by a kindergarten, a junior high school, a college and a university. I try to make objective observations of Japanese education from the real educational field. In the college and the university, the students make efforts to master each professional skill for preschool teacher, elementary school teacher, cooking license, fashion designer or tax accountant and so on. Although most students are not good at using practical English. In that context, the college and the university urgently promote the development of students' English skills.

In order to foster of human resources for Global Citizenship in Japan, it is necessary to educate young students with utilizing characteristics of the Japanese. Globalization is not a lack of individuality. Moreover, it is necessary to prevent young students from Indifference to other countries. I always encourage as many students as possible to form a partnership for Global Citizenship. Some of the students have become conscious of how global issues are done. I take it as my challenge to find and contribute to the fostering Global Citizenship gradually while supporting the students to develop their own freedom of expression.

Name	Sanjana Manaktala
Affiliation	Consultant, United Nations Information Centre for India and Bhutan
Nationality	Republic of India
Courses	MSc, UNU-ISP Postgraduate Degree Programme
Year	2012

Decent Work for All
Mobilising Youth Network Partnerships in India

Today, there are 1.8 billion people between the ages of 15 and 29, of which 87% live in developing countries. 41% of India's 1.3 billion people are below the age of 20. By the end of the century – the timeline of the Paris Climate Change Agreement – the adolescents of today will be the international leaders of tomorrow. But, in India, only 4.5% of the population is educated till the graduate level, and only 32.6% till the primary level. When correlated with youth unemployment numbers (around 13% globally and in India) a worrying scenario emerges. Along with negative implications for wellbeing and productivity, unemployed youth are also easily radicalized. These numbers suggest not a demographic dividend, but a ticking time bomb.

The question then becomes: How do we engage and equip young people to be peaceful and productive global citizens?

The challenge lies not only with job creation, but with skill development and employability. Creatively partnering with youth networks around the Sustainable Development Goals opens young people up to a world of opportunities for self-improvement and up-skilling.

The first step must be to create contact with the Goals. Through the UN Information Centre, school networks across the country have been introduced to the SDGs with activities including lectures, art, music, games and essays. UNIC has partnered with entities as diverse as CSR departments, chambers of commerce and animation schools, to conduct art exhibitions, theatre competitions and film awards. Activities like these contribute to the holistic development of a child, enhancing critical thinking, communication, creativity and social responsibility. But the net must widen to include deeper, more impactful work to catalyze the next step: initiating action.

UNIC ran a climate-themed science competition for almost 700 government schools in Tamil Nadu, replicated in Delhi with 30 private schools, where the winners were connected with college-level engineers to explore real-world implementation of the model. By assisting with research and writing for a book on the 70-year history of India and the UN, nearly twenty young interns have left UNIC after a powerful learning experience, with a richer, deeper commitment to the UN and the SDGs.

Activities aimed at young people must have a visible and tangible impact – on the outside world and on the participants themselves. Along with artistic and cultural activities, there must be large-scale measurable campaigns that include both physical and digital outreach. There is great potential in networks like Model UN participants, which can connect motivated young people across the globe with opportunities to develop initiatives around the SDGs. At the same time, there are also countless projects, enterprises and non-profits that already share the aims of the wide-ranging and ambitious SDG agenda. It is up to nodal organizations and individuals to draw these connections, bridge the gaps between global vision and local action, and broaden the scope of work under the SDGs.

Only by providing young people with opportunities to develop their capabilities can we reap the demographic dividend and create a safer, happier and more prosperous world for all.

Name	Ana Karen Mora Gonzalez
Affiliation	Chief Sustainability Officer, Fibra Uno
Nationality	United Mexican States (Mexico)
Courses	MSc, UNU-IAS Postgraduate Degree Programme (Yokohama)
Year	2012

Revitalize the Global Partnership for Sustainable Development
Payment for Ecosystem Services in Southern Mexico City

To achieve sustainable development, collaboration is imperative. A major challenge to achieve sustainable development is the lack of collaboration; amongst nations, private companies, public sector, NGO's and communities. Partnerships at all levels are required to successfully tackle gender inequality, environmental conservation, poverty, hunger, human rights protection, or any other current world problem.

Partnerships have proven to be the most effective solution for global issues at the local level. Working separately may bring results, but it typically takes longer periods of time, higher costs and additional resources. When different players come together into partnerships, and bring to the table their strengths, it delivers comprehensive results sooner.

While working at ICA, the largest construction company in Mexico and one of the largest in Latin America, we created a partnership amongst the Mexican Environmental Secretariat (SEMARNAT), the San Miguel and the Santo Tomás Ajusco (communities) and the company (ICA), to fund and pay for ecosystem services their forests provided to Mexico City.

Ostensibly it looks like a simple and basic project, however, it is important to consider the conditions it was embedded in. Mexico City was constructed on a lake, which has been drying up due to excessive water extraction, causing the city to sink. Given the need to maintain water filtration and aquifer recharge, the city declared 59% of its territory as conservation soil (SEDEMA, 2013) creating conflicting interests amongst farmers living in these conservation areas, who naturally, wanted to use and exploit the resources in their ejido-owned land.

Simultaneously, ICA as part of its environmental plan had been conducting reforestation campaigns for more than 60 years all over the country, but realized it needed community compromise, in order for these reforested areas to endure and deliver the ecosystem services they were intended for.

Moreover, we realized these campaigns required government cooperation to oversee the project in the long term and place it within a wider national environmental conservation strategy. Consequently, we brought ICA and SEMARNAT together, creating a concurrent fund to pay 2,345,000 MXN to the communities for the preservation, protection and sustainable use of 220Ha. Establishing the highest amount paid for conservation per hectare in the country, while creating the first public-private partnership for ecosystem services payments in Mexico City (SEMARNAT, 2012).

Initially the project was successful, it achieved the objectives of carbon sequestration, biodiversity conservation, support water permeation into the subsoil and community funding; but it did not drive community empowerment nor self-sufficiency.

Three years later, the program required restructuring to avoid being assistentialist and transform itself into an empowerment tool for the sustainable use of their resources. Currently the communities are diversifying their income from other sources such as Christmas trees production and apiculture, which hopefully will ensure long term local development and self-sufficiency.

In summary, this project's major achievement was, bringing together efforts the government and the private sector where doing separately. In three years both institutions, achieved what they had work for more than 60 years: The Conservation of ecosystem services while empowering communities in Mexico City.

Reference
- SEDEMA (Secretaría de Medio Ambiente de la Ciudad de México). 2013. Primer Informe SEDEMA. Capítulo 3 Suelo de Conservación y Biodiversidad. Disponible en: http://data.sedema.cdmx.gob.mx/sedema/images/archivos/noticias/primer-informe-sedema/capitulo-03.pdf
- SEMARNAT (Secretaría de Medio Ambiente y Recursos Naturales). 2012. Comunicado de prensa Núm. 29/12. Disponible en: http://tracsablog.typepad.com/files/ica-conafor.pdf

	Name	Elvira Tamayo Okabe
	Affiliation	Executive Director/Founder, Center for Elevating Others Global Foundation Director/Board Secretary, Raytech Dataprocessing Corporation
	Nationality	Republic of Philippines
	Courses	UNU Global Seminar – Okinawa Session
	Year	2003

General Solutions to Overcoming the Plight of the Poor and Eradicating Corruption All Over the World Based on Philippines' Situation

Eliminating Poverty and Corruption

1 Problem Statement

The use and abuse of the poor, in general:
- Human trafficking and rape cases;
- Unemployment, discrimination, low salary or compensation inequality;
- Illegal use of drugs through rich drug lords; and
- Lack of staple food subsidies, financial assistance to start Small and Medium Enterprises (SMEs), college educational assistance, free education until senior high school, health and insurance, physically and mentally handicapped benefits, senior citizen's benefits, and other emergency and calamity assistances.

Improper use of funds:
- Red tapes in distribution of goods and monetary assistance in times of emergency and in awarding micro-financial assistance;
- Lack of self-help projects and workshops at community level to encourage the poor people to be enterprising and independent in life.

2 Causal Analysis
- Poverty in poor countries and even in developing countries like the Philippines is caused by government corruption, red tapes, and social inequality.
- By eradicating poverty and providing free good education for all, people will have a more disciplined culture that will have ripple

effects in solving human trafficking, human abuse, drug use and trafficking, killing, and other inhumane acts.
- The end to poverty and corruption can be completely eliminated if all benefactors, donors, and beneficiaries could be disciplined enough to form a universal culture of equality, honesty and transparency without any vested interest in any transactions with among the animators and actors.

3 Proposed Solutions

- How to get rid of corruption, red tape or bureaucratic habits in the government and private sectors:

    a. The United Nations Organization (UNO) and attached agencies to train all concerned government agencies and non-government organizations (NGOs) of all member nations on all policies, procedures, and sanctions in order to completely eliminate corruptions and red tapes as far as helping the poor is concerned;

    b. Intensive and regular monitoring, coordination, and auditing of existing projects using measurements of leadership and teamwork accountability and responsibility for all individuals and groups; and

    c. The United Nations Organization (UNO) and attached agencies in every member country to directly make assessments of situations and direct deployment of assistances to target beneficiaries.

- Poor beneficiaries to be trained how to make good use of all available resources received at optimum level using economies of scale.
- UNO to give annual awards like the Nobel Prize to outstanding individuals and groups up to country level in order to encourage everyone to do their best as far as eradicating poverty and corruption are concerned.
- UNO to require all member nations, rich or poor, to make it a uniform policy to provide the poor with staple food subsidies, financial assistance to start Small and Medium Enterprises (SMEs), college educational assistance, free education until senior high school, health and insurance, physically and mentally handicapped benefits, senior citizen's benefits, and other emergency and calamity assistances.

Name	Shuan SadreGhazi
Affiliation	Research fellow, Innovation for Development, UNU-MERIT, Maastricht, the Netherlands
Nationality	Islamic Republic of Iran
Courses	UNU International Courses
Year	2007

New Modes of Partnership for Poverty Alleviation
Fostering the Link between the Private Sector and Social Entrepreneurs

In the last 4 decades, many development and aid initiatives have tried to deliver solutions for those living in poverty, involving mainly non-profit or public actors. They have had different levels of success and impacts. On the other hand, emergence and expansion of global challenges such as climate change, terrorism, infectious diseases and poverty, which cannot only be solved by the conventional international aid alone, led to exploration of new forms of international cooperation. In line with that, we witness increased involvement of private sector and social entrepreneurs in development initiatives.

In the last 10 years, my focus has been on innovations that benefit under-served communities. I have worked with a number of pro-poor initiatives and a variety of organizations and groups, from Japan, Holland and Germany, to India, Myanmar and Iran. One of the initiatives, which was funded by Sasakwa Peace Foundation, aimed at incubating technologies that could benefit underserved communities in Asia.

One related project was in Myanmar, where the majority of rural areas don't have access to electricity. There were Japanese companies who had promising solar technologies for off-grid electricity generation, but those companies did not have any experience in deploying their technologies in a rural, developing-country context and had no knowledge about the social context. So one missing part of the puzzle for clean energy in the South was the link between the technology providers and the local stakeholders.

At the same time, there were some local NGOs and Social

Entrepreneurs in Myanmar, that had valuable experience in dealing with the problems facing rural communities. They had the trust of the local communities, knew how to communicate with them and could address issues related to distribution and introduction of solar solutions. The social entrepreneurs were familiar with the social fabric of the community, but they lacked the technological know-how and resources to bring solar solutions to scale. This is where the partnership between the private sector -with technological know-how and resources- on one hand , and the input of social entrepreneurs – with knowledge of the local context- on the other hand, became instrumental.

Collaboration of the Japanese private sector with the Myanmar social entrepreneurs for green energy access, proved to be a successful case in the new modes of international development initiatives (A number of similar technology-based projects in the past had failed, as they did not consider the local/social aspects and only aimed at distributing the technology). A similar situation was for sanitation in rural India, where again the development impact of the projects that worked effectively with the social entrepreneurs was much higher.

I believe both private sector and social entrepreneurs have vast potential in addressing the challenges of delivering effective solutions to address SDGs in the South. However, each group has its own strengths and weaknesses. Their potentials in delivering inclusive innovations will complement and materialize only if they manage to build an effective and sustained partnership. Policy makers and international donors can have a pivotal role in incentivizing and facilitating such partnerships.

Name	Naushad Ali Yakub
Affiliation	UNU Alumni Association
Nationality	Republic of Fiji
Courses	UNU-ISP Building Resilience to Climate Change (BRCC)
Year	2012

Partnership for Global Citizenship on SDG 14
Fiji, an Exemplar in Integrated Ocean Management: Supporting Partnership for Global Citizenship

Fiji, located in Oceania; is a developing country, with pressures from increasing population, geographic isolation from major employment centers, low wage rates, high socio-economic expenditure, and demands from cultural and religious activities; are posing additional pressure on natural resources to sustain livelihoods. As a result, natural resources are seen as a 'quick' solution to meet livelihood demands without prioritizing sustainability and conservation, despite included in much legislation, policies, strategies and development plans. Dual governance system is another key challenge between *rights to using resources* and *rights to own resources*; for communities, developers and conservationists. Now working for International Union for Conservation of Nature, Oceania Regional Office (IUCN-ORO) in providing support to Fiji Government to establish 30% network of marine protected areas (MPAs) for its seas by 2020 to achieve Aichi Target 11 under Convention on Biological Diversity (CBD) and Sustainable Development Goals (SDG) 14. Specifically, the Marine and Coastal Biodiversity Management in Pacific Island Countries (MACBIO) Project funded by International Climate Initiative (IKI) of the German Federal Ministry for Environment, Nature Conservation and Nuclear Safety (BMUB). It is being implemented by the German Agency for International Cooperation (GIZ) together with the Fiji Government in close collaboration with the Secretariat of the Pacific Regional Environment Programme (SPREP) and with technical support from the International Union for Conservation of Nature (IUCN). The objective of the project "Strengthening the sustainable management of marine and coastal biodiversity in Fiji." This Project is the Secretariat of the Marine Working Group under Protected Areas Committee that is formed under Environment Management Act and Fiji's MPA Technical Committee under the Offshore Fisheries Management Decree. These committees are comprised of NGOs and Government agencies and shows national partnership. MACBIO project supports Fiji's commitments to International, Regional and National obligations such as CBD, Convention on Migratory Species (CMS), Convention on International Trade on Endangered Species (CITIES), Samoa Pathways, Western Central Pacific Fisheries Commission, Endangered and Protected Species Act (Fiji 2002), Environment Management Act (Fiji 2005), Offshore Fisheries Management Decree (Fiji 2012), Green Growth Framework (Fiji 2014) and National Development Plan (Fiji 2015). Marine Spatial Planning is currently used to suggest and design a network of MPAs. This exercise incorporates recommendations from legal review pertaining to MPAs, identifying special and unique marine areas, possible MPA types and describing value neutral marine environment of Fiji (bioregions) on candidate sites; are innovative ways in which marine conservation can be achieved. This support displays Regional Oceania Partnership towards Global Citizenship. The vision for Fiji's network of marine protected areas is: *A comprehensive, ecologically representative network of MPAs that restores and sustains the health, productivity, resilience, biological diversity and ecosystem services of coastal and marine systems, and promote the quality of community's livelihoods.* These collaborative efforts nationally, regionally and internationally clearly display support for global citizenship through partnership.

Name	Wendy Mei Tien Yee
Affiliation	University Malaya
Nationality	Malaysia
Courses	UNU International Courses
Year	2006

Quality Education: Education for Peace and Humanities in Malaysia

Creating Solidarity, Restoring Hope and Empowerment of Youth by 2030: Education for Peace and Humanities in Malaysia to Promote the Core Values of Respecting Life's Inherent Dignity

1 Problem Statement

In Malaysia, although majority of the population has access to education and fulfilled the compulsory education requirement set forth by the Millennium Development Goals, there are still issues regarding the quality of education the young people are receiving and the lifelong learning opportunities for all Malaysians.

2 Causal Analysis

- The curriculum at the school is constantly changing. Teachers and students were not prepared for the sudden changes and thus affect the quality of the teaching and learning processes.
- There is a huge gap between the national education blue print with the actual implementation of education at the schools. As a result, it affected the lifelong learning mentality of the students.
- There is an overemphasis on examination at school. Hence, students who performed better are given more attention while students who are weaker in academic performance are given lesser attention. Hence, Malaysia faces a serious problem with equitable education for all.
- There are still many cases of bullies, gangsterism, drug trafficking at schools. Many students who committed such social ills are mainly from the neglected pool of academically weak students. As a result, many of these students will eventually drop out without completing their high school and the vicious cycle of a failed education continues in the society.

3 Proposing Action to be Supported
- A pilot project entitled 'Education for Peace and Humanities in Malaysia' will be developed to address the above problems.
- In this project, it will include the following action plans:
  - ➢ To develop a comprehensive educational curriculum for the teachers' training at the university. This curriculum focuses on the core values of education that is character building and respecting the dignity of all life. This curriculum also includes the development of teachers' own character as the teachers will act as role models in school.
  - ➢ To develop an education policy that moves away from exam based learning to student development learning. Students' assessments are based on their individual capability and their individual improvements instead of a standard yardstick. This policy aims to cultivate lifelong learning and allow every student to develop and grow according to their individual potential; thus achieving an equitable education for all.
  - ➢ To form Peace Clubs, Human Rights Clubs or United Nations Club at the schools. These after school programs will encourage students to be involved in value creating activities and develop social skills which will enable them to be socially competent to curb them from committing social ills.

あとがき

　この本は、国連機関の一つである国連大学（United Nations University: UNU）と公益財団法人国連大学協力会（The Japan Foundation for the United Nations University: jfUNU）が、2017年3月11日〜12日に東京で開催した国際シンポジウム「持続可能な地球社会を目指して‐私のSDGsへの取り組み」における議論の結果を、その後に得られた新たな知見をも加えて、シンポジウムの参加者たちが著者となって書いたものである。

　この本の特徴は、国連が2030年までに全地球規模で達成したいとして設定した持続可能な開発目標（Sustainable Development Goals: SDGs）に関し、国連大学の修了生たち個々人がどう取り組んでいるかを発表し合ったうえで、専門家を交えて議論をしている点にある。つまり、個々人がSDGsにどう向き合うかを普段の実践の中から議論していることに大きな特徴がある。

　jfUNUは、1985年に、文部大臣経験を持つ永井道雄氏が国連大学を支援するために産官学の幅広い有志の方々に呼びかけて設立した民間団体である。jfUNUは、その後多様な形で国連大学を支援してきたが、2007年に国連大学の修了生たちによる同窓会の事務局を引き受けることとした。国連大学の修了生たちは、この同窓会を通してつながり、国連大学を離れた今

も、国連や世界が直面する緊急課題に取り組み、情報交換や交流をしている。

　このシンポジウムにはそのようにして育てられた若者たちが、世界各地、SDGsの多様な分野から参集していることに大きな意味がある。このシンポジウムの運営には、永井道雄氏の遺志によりjfUNU内に設定された永井基金が充てられ、また、一般社団法人東京倶楽部からも助成を得た。

　この本を読んでいただいた多くの方々が、SDGsへの理解と関心を高め、国連大学の学生たちへの活動に温かい支援の目を寄せていただくことを願って、ここに記して、心からの謝意としたい。

　　　　　　　　　　　　　　　公益財団法人国連大学協力会
　　　　　　　　　　　　　　　常務理事・事務局長　森　茜

付　録

持続可能な開発目標（SDGs）

目標1：あらゆる場所で、あらゆる形態の貧困に終止符を打つ

目標2：飢餓に終止符を打ち、食料の安定確保と栄養状態の改善を達成するとともに、持続可能な農業を推進する

目標3：あらゆる年齢のすべての人々の健康的な生活を確保し、福祉を推進する

目標4：すべての人々に包摂的かつ公平で質の高い教育を提供し、生涯学習の機会を促進する

目標5：ジェンダーの平等を達成し、すべての女性と女児のエンパワーメントを図る

目標6：すべての人々に水と衛生へのアクセスと持続可能な管理を確保する

目標7：すべての人々に手ごろで信頼でき、持続可能かつ近代的なエネルギーへのアクセスを確保する

目標8：すべての人々のための持続的、包摂的かつ持続可能な経済成長、生産的な完全雇用およびディーセント・ワークを推進する

目標9：レジリエントなインフラを整備し、包摂的で持続可能な産業化を推進するとともに、イノベーションの拡大を図る

目標10：国内および国家間の不平等を是正する

付　録　203

　目標11：都市と人間の居住地を包摂的、安全、レジリエントかつ持続可能にする

　目標12：持続可能な消費と生産のパターンを確保する

　目標13：気候変動とその影響に立ち向かうため、緊急対策を取る

　目標14：海洋と海洋資源を持続可能な開発に向けて保全し、持続可能な形で利用する

　目標15：陸上生態系の保護、回復および持続可能な利用の推進、森林の持続可能な管理、砂漠化への対処、土地劣化の阻止および逆転、ならびに生物多様性損失の阻止を図る

目標16：持続可能な開発に向けて平和で包摂的な社会を推進し、すべての人々に司法へのアクセスを提供するとともに、あらゆるレベルにおいて効果的で責任ある包摂的な制度を構築する

目標17：持続可能な開発に向けて実施手段を強化し、グローバル・パートナーシップを活性化する

〔出典：国際連合広報センターホームページ〕

＜執筆者・編者紹介＞

沖　大幹

国際連合大学上級副学長、国際連合事務次長補。

東京大学工学部卒業、博士（工学、1993 年、東京大学）。東京大学生産技術研究所助教授、文部科学省大学共同利用機関・総合地球環境学研究所助教授などを経て、2006 年東京大学教授。2016 年 10 月より現職。現在も東京大学総長特別参与、教授（国際高等研究所サステイナビリティ学連携研究機構）を兼任。

地球規模の水文学および世界の水資源の持続可能性を研究。気候変動に関わる政府間パネル（IPCC）第 5 次報告書統括執筆責任者、国土審議会委員ほかを務める。生態学琵琶湖賞、日経地球環境技術賞、日本学士院学術奨励賞など表彰多数。水文学部門で日本人初のアメリカ地球物理学連合（AGU）フェロー（2014 年）。

書籍に『水の未来 ― グローバルリスクと日本 』（岩波新書、2016 年）、『水危機 ほんとうの話』（新潮選書、2012 年）、『水の世界地図第 2 版』（監訳、丸善出版、2011 年）、『東大教授』（新潮新書、2014 年）など

勝間　靖

早稲田大学大学院アジア太平洋研究科（国際関係学専攻）教授、国立国際医療研究センター・グローバルヘルス政策研究センター国際保健外交・ガバナンス研究科長。

UNU グローバルセミナー　1988 年および 1989 年修了。

国際基督教大学教養学部および大阪大学法学部を卒業後、同大学院で法学修士号、ウィスコンシン大学マディソン校で Ph.D.（開発学）を取得。国連児童基金（UNICEF）職員を経て、現職。国連開発計画（UNDP）で『人間開発報告書』諮問委員、英国の医学誌『BMJ』で諮問委員、

国際開発学会で理事、国際人権法学会で理事を務める。過去に、外務省独立行政法人評価委員、ジョージワシントン大学エリオット国際情勢スクール客員研究員、国際開発学会副会長、日本国際連合学会事務局長を務めた。

専門分野：開発研究（人間開発）、国際人権論（子どもの権利）、人間の安全保障、グローバル・ガバナンス

主な編著書：『テキスト国際開発論〜貧困をなくすミレニアム開発目標へのアプローチ』（ミネルヴァ書房、2012）、『国際社会を学ぶ』（晃洋書房、2012）、『アジアの人権ガバナンス』（勁草書房、2011）、『国際緊急人道支援』（ナカニシヤ出版、2008）など

弓削昭子

法政大学法学部国際政治学科教授・前 UNDP 駐日代表・総裁特別顧問。米国コロンビア大学教養学部卒。ニューヨーク大学大学院で開発経済学修士号取得。国連開発計画（UNDP）タイ事務所で勤務を始め、ニューヨーク UNDP 本部に転勤。1983 年に帰国、社団法人海外コンサルティング企業協会で勤務後、開発コンサルタントとして活動。1988 年に UNDP タイ事務所常駐代表補佐、1990 年 UNDP インドネシア事務所常駐副代表、1994 〜 98 年 UNDP ブータン事務所常駐代表。1999 〜 2002 年フェリス女学院大学国際交流学部教授。2002 年から UNDP 駐日代表を務め、2006 年に国連事務次長補・国連開発計画（UNDP）管理局長就任。2012 〜 13 年、UNDP 駐日代表・総裁特別顧問を務めた。2014 年 4 月より現職。

滝澤美佐子

桜美林大学大学院国際学研究科教授。
UNU グローバルセミナー　1987 年修了。

津田塾大学学芸学部国際関係学科卒業。国際基督教大学大学院修了・博士（学術）。ロンドン大学国際法ディプロマコース修了。中部大学、同大学院を経て 2005 年より現職。日本国際連合学会理事。日本財団ハンセン病差別撤廃原則とガイドラインフォローアップ事業国内アドバイザリー委員会委員（2012-14）。一橋大学・ロンドンスクールオブエコノミクス客員研究員（2014-15）。

専門分野：国際法、国際機構論、国際人権法

主な著書：『国際人権基準の法的性格』（国際書院、2004 年）、共著に『国際人権入門』（法律文化社、2008 年、第 2 版 2013 年）『国際社会と法』（有斐閣、2010 年）、『コンメンタール女性差別撤廃条約』（尚学社、2010 年）『国際機構論　総合編』（国際書院、2016 年）、『入門国際機構』（法律文化社、2016 年）など

吉高神　明

福島大学経済経営学類教授。

UNU グローバルセミナー　1985 年（第 1 回）修了。

青山学院大学国際政治経済学部卒業。福島大学経済学部助教授、同経済経営学類准教授を経て、2008 年 4 月より現職。現在、福島県労働委員会公益委員、放送大学（福島学習センター）客員教授、会津大学非常勤講師等を兼務。

専門分野：国際公共政策論

主な著書：『環境と開発の国際政治』（共著　南窓社　1999 年）、『東日本大震災からの復旧・復興と国際比較』（共著　八朔社　2014 年）など

杉村美紀

上智大学グローバル化推進担当副学長、総合人間科学部教育学科教授。
UNU グローバルセミナー　1987 年修了。

博士（教育学・東京大学）外務省専門調査員、国立教育研究所（現、国立教育政策研究所）研究協力者、広島大学教育開発国際協力センター客員研究員を経て 2002 年より上智大学に勤務。国連大学協力会助成諮問委員会委員（2011 年〜現在）、日本学生支援機構留学生交流事業実施委員会委員（2013 年〜現在）、日中友好会館評議員（2016 年〜現在）、ユネスコ国内委員会委員（2016 年〜現在）、日本比較教育学会会長（2017 年〜現在）等を務める。

専門分野：比較教育学、国際教育学

主な著書：『移動する人々と国民国家：ポスト・グローバル化時代における市民社会の変容』（明石書店、2017 年、編著）『人間の安全保障と平和構築』（日本評論社、2017 年、共著）『多文化共生社会におけるＥＳＤ・市民教育』（上智大学出版、2014 年、共編著）『トランスナショナル高等教育の国際比較』（東信堂、2014 年、共著）など

齊藤　修

国連大学サステイナビリティ高等研究所・研究科長 / 学術研究官。米国タフツ大学大学院都市・環境政策研究科修士課程を修了。2004 年東京農工大学大学院連合農学研究科から博士（農学）学位を取得。大阪大学大学院工学研究科と早稲田大学高等研究所での助教を経て、2011 年 1 月から現職。2011 年 4 月から東京大学の客員准教授を兼務。2011 年からはサステイナビリティ・サイエンス誌（英文国際誌）の編集幹事を兼務。

専門分野：ランドスケープ・エコロジー、環境システム研究

主な著書：『工学生のための基礎生態学』（2017 年、理工図書、共著）『農村計画と生態系サービス』（2014 年、農林統計出版、共著）『里山・里海：自然の恵みと人々の暮らし』（2012 年、朝倉書店、共著）』

秋月弘子

亜細亜大学国際関係学部教授。

国際基督教大学大学院行政学研究科博士課程修了（学術博士）。1987 - 1989 年、国連開発計画（UNDP）プログラム・オフィサー、2002 年より現職、2005 - 2006 年、コロンビア大学客員研究員。日本国際連合学会理事・広報主任、東京オリンピック・パラリンピック組織委員会「持続可能な調達ワーキンググループ」座長、第 7 次出入国管理政策懇談会委員等を兼務。

専門分野：国際法、国連研究

主な著書：『国連法序説』（単著、国際書院、1999 年）、『国際機構論総合編』（共著、国際書院 2015 年）、『国際社会における法と裁判』（共著、国際書院、2014 年）、『人類の道しるべとしての国際法』（共著、国際書院、2011 年）、『国際法入門』（共著、有斐閣、2005 年）など

森　茜

公益財団法人国連大学協力会常務理事兼事務局長。大学卒業、文部科学省にて学術行政・教育行政に携わるとともに、内閣府にて青少年問題・婦人問題に携わる。その後、複数の国立大学の管理職を務めた後、図書館情報大学事務局長、駿河台大学理事兼事務局長を経て、現職。

主な活動分野は高等教育行政、学術行政、大学管理運営。

なお、2013 年以来、公益社団法人日本図書館協会理事長。図書館情報学に精通し、いくつかの大学で図書館学の非常勤講師を行うとともに、図書館学の著作を有する。

索引

Addis Ababa Action Agenda 40, 51
CECAR-Africa 92
ECOSOC 29
ESG投資 100
Global Gender Gap Index 56
global partnerships 53
Global Sustainable Development Report 45
Goal 10 50
Goal 13 of SDGs 167
Goal 16 50
goal 7 of SDGs 169
goal number 3. 146
High-Level Political Forum on Sustainable Development (HLPF) 33, 45
IHR 122
Inter-Agency and Expert Group on Sustainable Development Goal Indicators. 49
MAPS approach 33, 54
New Urban Agenda 41
OCHA 123
Official Development Assistance (ODA) 36, 51, 56
Paris Agreement 41
SDG 4 151
SDG 5 154
SDG 6 170
SDG 7 170
SDG 8 170
SDG 9 170
SDG 11 150, 170
SDG 12 170
SDG 13 170
SDG 14 192
SDG 15 150
SDG Goal 7 157
SDGs Implementation Guiding Principles 55, 56
SDGsアクションプラン2018 98
SDG Target 3.8 136
SDG実験室（SDG Labs） 95
Sustainable Development Goals（SDG）14. 192
Sustainable Development Goals Report 45
the SDGs Promotion Headquar-

ters　54
The Sendai Framework for Disaster Risk Reduction　40
the United Nations Conference on Housing and Sustainable Urban Development (known as HABITAT III)　41
the United Nations Development Group (UNDG)　53
Third International Conference on Financing for Development.　40
UNHCR　114
United Nations Economic and Social Council (ECOSOC).　45
United Nations Statistical Commission　49, 50
United Nations Summit on Addressing Large Movements of Refugees and Migrants　41
UNMEER　118, 120, 122
UNRWA　114
Voluntary National Review (VNR)：45
WHO　118
World Humanitarian Summit　41

あ行

アディス・アベバ行動目標　18, 30, 65
安全保障理事会　120

い行

イスラム教　110
一般社団法人日本経済団体連合会（経団連）　99
インターリンケージ　91, 94

え行

エボラ出血熱　116

お行

欧州連合 (European Union: EU)　85

か行

開発資金国際会議　18
学際的　85, 122
環境・社会・企業統治 (Environmental, Social and Corporate Governance: ESG)　99

き行

企業行動憲章　99

気候変動枠組条約（United Nations Framework Convention on Climate Change: UNFCCC） 65
共生 81, 83
強制移住 110
キリスト教 111
緊急対応枠組み（Emergency Response Framework: ERF） 118

く行

グローバル・ガバナンス 64, 68, 107, 123
グローバル・ジェンダー・ギャップ指数（Global Gender Gap Index） 36
グローバル・シティズンシップ 81, 82
グローバル・シティズンシップ教育（Global Citizenship Education: GCED）」 83
グローバル・パートナーシップ 32, 64, 66, 67, 69
グローバル社会 101, 107
グローバルヘルス・ガバナンス 117, 121

け行

経済協力開発機構（Organisation for Economic Cooperation and Development: OECD） 97

こ行

国際科学会議（The International Council for Science: ICSU） 91
国際協力機構（Japan International Cooperation Agency: JICA） 114
国際人権法 111
国際人道法 111
国際的に懸念される公衆衛生上の緊急事態（Public Health Emergency of International Concern: PHEIC） 116
国際保健規則（International Health Regulations: IHR） 118
国内行動計画（National Action Plan: NAP） 68
国民総所得（Gross National Income: GNI） 67
国民総生産（Gross Domestic Product: GDP） 93
国連エボラ緊急対応ミッション

(United Nations Mission for Ebola Emergency Response: UNMEER) 117
国連開発援助枠組み (United Nations Development Assistance Framework: UNDAF) 122
国連開発グループ (United Nations Development Group) 33
国連環境計画 (United Nations Environment Programmes: UNEP) 93, 99
国連気候変動枠組条約第21回締約国会議 (The 21st Conference of the Parties to the United Nations Framework Convention on Climate Change: COP21) 19, 74
国連グローバル・コンパクト (United Nations Global Compact: UNGC) 99
国連経済社会理事会 (United Nations Economic and Social Council: ECOSOC) 22
国連児童基金 (United Nations Children's Fund: UNICEF) 115
国連持続可能な開発のための教育の10年 (United Nations Decade of Education for Sustainable Development: DESD) 84
国連人道問題調整事務所 (United Nations Office for Coordination of Humanitarian Affairs: OCHA) 121
国連大学 (United Nations University: UNU) 85, 89
国連大学サステイナビリティ高等研究所 (United Nations University Institute for the Advanced Study of Sustainability: UNU-IAS) 90
国連中央緊急対応基金 (Central Emergency Response Fund: CERF) 121
国連統計委員会 (United Nations Statistical Commission) 28
国連統計部 29
国連難民高等弁務官事務所 (United Nations High Commissioner for Refugees: UNHCR) 109
国連人間居住会議 19
国連パレスチナ難民救済事業機関 (United Nations Relief and Works Agency for Palestine

Refugees in the Near East: UNRWA) 112
国連平和維持活動 (Peace Keeping Operations: PKO) 114
国連防災世界会議 18, 73
国境なき医師団 117, 118, 120
子どもの権利条約 65, 111

し 行

自然資本 90, 93
持続可能性に配慮した調達コード 98, 100
持続可能な開発グローバル報告書 24, 68
持続可能な開発サミット 18, 73
持続可能な開発に関するハイレベル政治フォーラム (High-Level Political Forum on Sustainable Development: HLPF) 22
持続可能な開発のための教育 (Education for Sustainable Development: ESD) 82, 83
持続可能な開発目標 (SDGs) 実施指針 34, 98
持続可能な開発目標 (SDGs) 指標に関する機関間専門家グループ (Inter-Agency and Expert Group on Sustainable Development Goal Indicators) 28
持続可能な開発目標 (SDGs) 推進本部 34, 68, 98
持続可能な開発目標報告書 23, 68
持続可能な共生社会 86
持続可能な社会 81, 90, 99, 100
自発的国家レビュー (Voluntary National Review: VNR) 22
社会的排除 108
シリア難民 112
人権侵害 109, 110, 113
新興感染症 108, 122
人工資本 93
人的資本 93
人道危機 110, 112, 121, 122

せ 行

生活習慣病対策手帳 115
政府開発援助 31, 67, 114
生物多様性 90, 98
世界教育フォーラム (World Education Forum: WEF) 82
世界首脳会議成果文書 (2005年) 65
世界人道サミット 19
世界保健機構 (World Health Organization: WHO) 116

責任投資原則　99, 100
仙台防災枠組　76
仙台防災枠組 2015-2030　18, 73, 75

た 行

第三国定住　109, 113
ダカール枠組み　82
多文化化　83
多文化共生　113
多様化　83, 92

ち 行

地球規模感染症に対する警戒と対応ネットワーク（Global Outbreak Alert and Response Network: GOARN）　117

て 行

低開発国　66

と 行

東南アジア諸国連合（Association of SouthEast Asian Nations: ASEAN）　85

な 行

難民と移民に関する国連サミット　19
難民と移民に関するニューヨーク宣言　109
難民に関するグローバル・コンパクト　109

に 行

ニュー・アーバン・アジェンダ　19
人間の安全保障　74, 109, 113, 116

は 行

パリ協定　19, 74
万人のための教育（Education for All: EFA）　82

ひ 行

ビジネスと人権に関する指導原則　99
兵庫行動枠組　73

ふ 行

フューチャー・アース　95
武力紛争　110, 121
文化的国際主義（Cultural Internationalism）　85, 86

へ行

平和構築　114

ほ行

包括的富指標　93, 94
ボート・ピープル　114
ボコ・ハラム　109, 110
母子健康手帳　115, 116

も行

目標1　75, 91
目標2　91
目標2.4　75
目標3　91, 107, 123
目標4　82
目標6　91
目標7　91
目標9　75, 90, 91
目標10　29
目標11　75, 91
目標11.b　75
目標11.c　75
目標13　65, 90
目標13.1　75
目標14　90, 91
目標14.2　75
目標15　90, 91
目標16　29
目標17　67

ゆ行

ユネスコ（国際連合教育科学文化機関、United Nations Educational, Scientific and Cultural Organization; UNESCO）　82
ユネスコスクール　84

れ行

レジリエンス　74, 75, 77, 90, 92, 94

ろ行

ロヒンギャ　111, 115

jfUNU レクチャー・シリーズについて

UNU（国連大学）は，日本に唯一本部をおく国際機関です．

UNU は，世界が直面する地球規模の緊急課題を調査研究し，世界に提言を行うと共に，それに携わる若い人材の育成を図っています．

jfUNU（国連大学協力会）は，そのような UNU の活動を支援するための公益財団法人です．この jfUNU レクチャー・シリーズは，UNU 及び jfUNU の活動の一端をシリーズとして刊行し，世界の緊急課題について，より多くの人々に知ってもらい，その解決に役立てていただくことを願ってお届けするものです．

jfUNU レクチャー・シリーズ ⑩

持続可能な地球社会をめざして：
わたしの SDGs への取組み

編者　勝間　靖

2018 年 9 月 2 日初版第 1 刷発行

・発行者──石井　彰
印刷・製本／モリモト印刷株式会社
ⓒ 2018 by Japan Foundation for
United Nations University
(jfUNU)
（定価＝本体価格 2,000 円＋税）
ISBN978-4-87791-292-5 C3032 Printed in Japan

・発行所

KOKUSAI SHOIN Co., Ltd.
3-32-6, HONGO, BUNKYO-KU, TOKYO, JAPAN.
株式会社 **国際書院**
〒113-0033 東京都文京区本郷 3-36-2-1001
TEL 03-5684-5803　　FAX 03-5684-2610
E メール：kokusai@aa.bcom.ne.jp
http://www.kokusai-shoin.co.jp

本書の内容の一部あるいは全部を無断で複写複製(コピー)することは法律でみとめられた場合を除き，著作者および出版社の権利の侵害となりますので，その場合にはあらかじめ小社あて許諾を求めてください．

国際法

東 壽太郎・松田幹夫編
国際社会における法と裁判

87791-263-5 C1032　　　　A5判 325頁 2,800円

尖閣諸島・竹島・北方領土問題などわが国を取り巻く諸課題解決に向けて、国際法に基づいた国際裁判は避けて通れない事態を迎えている。組織・機能・実際の判決例を示し、国際裁判の基本的知識を提供する。
(2014.11)

渡部茂己・望月康恵編著
国際機構論 [総合編]

87791-271-0 C1032　　　　A5判 331頁 2,800円

「総合編」、「活動編」「資料編」の3冊本として順次出版予定。「総合編」としての本書は、歴史的形成と発展、国際機構と国家の関係、国際機構の内部構成、国際機構の使命など第一線で活躍している専門家が詳述。
(2015.10)

波多野里望／松田幹夫編著
国際司法裁判所
—判決と意見第1巻（1946-63年）

906319-90-4 C3032　　　　A5判 487頁 6,400円

第1部判決、第2部勧告的意見の構成は第2巻と変わらず、付託事件リストから削除された事件についても裁判所年鑑や当事国の提出書類などを参考にして事件概要が分かるように記述されている。
(1999.2)

波多野里望／尾崎重義編著
国際司法裁判所
—判決と意見第2巻（1964-93年）

906319-65-7 C3032　　　　A5判 561頁 6,214円

判決及び勧告的意見の主文の紹介に主眼を置き、反対意見や分離（個別）意見は、必要に応じて言及する。事件概要、事実・判決・研究として各々の意見を紹介する。巻末に事件別裁判官名簿、総名簿を載せ読者の便宜を図る。
(1996.2)

波多野里望／廣部和也編著
国際司法裁判所
—判決と意見第3巻（1994-2004年）

87791-167-6 C3032　　　　A5判 621頁 8,000円

第二巻を承けて2004年までの判決および意見を集約し、解説を加えた。事件概要・事実・判決・主文・研究・参考文献という叙述はこれまでの形式を踏襲し、索引もまた読者の理解を助ける努力が施されている。
(2007.2)

横田洋三／廣部和也編著
国際司法裁判所
—判決と意見第4巻（2005-2010年）

87791-276-5 C3032　　　　A5判 519頁 6,000円

1999年刊行を開始し、いまや国際法研究者必読の書として親しまれている。第4巻は2005-2010年までの国際司法裁判所の判決および勧告的意見を取上げ、事件概要・事実・判決・研究を紹介する
(2016.8)

横田洋三／東壽太郎／森喜憲編著
国際司法裁判所
—判決と意見第5巻

87791-286-4 C3032　　　　A5判 539頁 6,000円

本書は2011-2016年までの国際司法裁判所が出した判決と勧告的意見の要約および開設を収録している。判決・勧告的意見の本文の紹介を主な目的とし、反対意見・分離意見は必要に応じて「研究」で言及した。
(2018.1)

横田洋三訳・編
国際社会における法の支配と市民生活

87791-182-9 C1032　　　　四六判 131頁 1,400円

［ jfUNUレクチャー・シリーズ①］ 東京の国際連合大学でおこなわれたシンポジウム「より良い世界に向かって−国際社会と法の支配」の記録である。本書は国際法、国際司法裁判所が市民の日常生活に深いかかわりがあることを知る機会を提供する。
(2008.3)

内田孟男編
平和と開発のための教育
—アジアの視点から

87791-205-5 C1032　　　　A5判 155頁 1,400円

［ jfUNUレクチャー・シリーズ②］ 地球規模の課題を調査研究、世界に提言し、それに携わる若い人材の育成に尽力する国連大学協力会（jfUNU）のレクチャー・シリーズ②はアジアの視点からの「平和と開発のための教育」
(2010.2)

国際法

井村秀文編
資源としての生物多様性
87791-211-6 C1032　　　　　　　A5判 181頁 1,400円

［*jf*UNU レクチャー・シリーズ③］気候変動枠組み条約との関連を視野にいれた「遺伝資源としての生物多様性」をさまざまな角度から論じており、地球の生態から人類が学ぶことの広さおよび深さを知らされる。
(2010.8)

加来恒壽編
グローバル化した保健と医療
——アジアの発展と疾病の変化
87791-222-2 C3032　　　　　　　A5判 177頁 1,400円

［*jf*UNU レクチャー・シリーズ④］地球規模で解決が求められている緊急課題である保健・医療の問題を実践的な視点から、地域における人々の生活と疾病・保健の現状に焦点を当て社会的な問題にも光を当てる。
(2011.11)

武内和彦・勝間 靖編
サステイナビリティと平和
——国連大学新大学院創設記念シンポジウム
87791-224-6 C3021　　　　　　　四六判 175頁 1,470円

［*jf*UNU レクチャー・シリーズ⑤］エネルギー問題、生物多様性、環境保護、国際法といった視点から、人間活動が生態系のなかで将来にわたって継続されることは、平和の実現と統一されていることを示唆する。
(2012.4)

武内和彦・佐土原聡編
持続可能性とリスクマネジメント
——地球環境・防災を融合したアプローチ
87791-240-6 C3032　　　　　　　四六判 203頁 2,000円

［*jf*UNU レクチャー・シリーズ⑥］生態系が持っている多機能性・回復力とともに、異常気象、東日本大震災・フクシマ原発事故など災害リスクの高まりを踏まえ、かつグローバル経済の進展をも考慮しつつ自然共生社会の方向性と課題を考える。
(2012.12)

武内和彦・中静 透編
震災復興と生態適応
——国連生物多様性の10年とRIO+20に向けて
87791-248-2 C1036　　　　　　　四六判 192頁 2,000円

［*jf*UNU レクチャーシリーズ⑦］三陸復興国立公園（仮称）の活かし方、生態適応の課題、地域資源経営、海と田からのグリーン復興プロジェクトなど、創造的復興を目指した提言を展開する。
(2013.8)

武内和彦・松隈潤編
人間の安全保障
——新たな展開を目指して
87791-254-3 C3031　　　　　　　A5判 133頁 2,000円

［*jf*UNU レクチャー・シリーズ⑧］人間の安全保障概念の国際法に与える影響をベースに、平和構築、自然災害、教育開発の視点から、市民社会を形成していく人間そのものに焦点を当てた人材を育てていく必要性を論ずる。
(2013.11)

武内和彦編
環境と平和
——より包括的なサステイナビリティを目指して
87791-261-1 C3036　　　　　　　四六判 153頁 2,000円

［*jf*UNU レクチャー・シリーズ⑨］「環境・開発」と「平和」を「未来共生」の観点から現在、地球上に存在する重大な課題を統合的に捉え、未来へバトンタッチするため人類と地球環境の持続可能性を総合的に探究する。
(2014.10)

日本国際連合学会編
21世紀における国連システムの役割と展望
87791-097-2 C3031　　　　　　　A5判 241頁 2,800円

［国連研究①］平和・人権・開発問題等における国連の果たす役割、最近の国連の動きと日本外交のゆくへなど「21世紀の世界における国連の役割と展望」を日本国際連合学会に集う研究者たちが縦横に提言する。
(2000.3)

日本国際連合学会編
人道的介入と国連
87791-106-5 C3031　　　　　　　A5判 265頁 2,800円

［国連研究②］ソマリア、ボスニア・ヘルツェゴビナ、東ティモールなどの事例研究を通じ、現代国際政治が変容する過程での「人道的介入」の可否、基準、法的評価などを論じ、国連の果たすべき役割そして改革と強化の可能性を探る。
(2001.3)

国際法

日本国際連合学会編
グローバル・アクターとしての国連事務局
87791-115-4　C3032　　A5判　315頁　2,800円

[国連研究③] 国連システム内で勤務経験を持つ専門家の論文と、研究者としてシステムの外から観察した論文によって、国際公務員制度の辿ってきた道筋を振り返り、国連事務局が直面する数々の挑戦と課題とに光を当てる。　　　　　(2001.5)

日本国際連合学会編
国際社会の新たな脅威と国連
87791-125-1　C1032　　A5判　281頁　2,800円

[国連研究④] 国際社会の新たな脅威と武力による対応を巡って、「人間の安全保障」を確保する上で今日、国際法を実現するために国際連合の果たすべき役割を本書では、様々な角度から追究・検討する。　　　　　　　　　　(2003.5)

日本国際連合学会編
民主化と国連
87791-135-9　C3032　　A5判　344頁　3,200円

[国連研究⑤] 国連を初めとした国際組織と加盟国の内・外における民主化問題について、国際連合および国際組織の将来展望を見据えながら、歴史的、理論的に、さらに現場の眼から考察し、改めて「国際民主義」を追究する。　　(2004.5)

日本国際連合学会編
市民社会と国連
87791-147-2　C3032　　A5判　311頁　3,200円

[国連研究⑥] 本書では、21世紀市民社会の要求を実現するため、主権国家、国際機構、市民社会が建設的な対話を進め、知的資源を提供し合い、よりよい国際社会を築いていく上での知的作用が展開される。　　　　　　　　　(2005.5)

日本国際連合学会編
持続可能な開発の新展開
87791-159-6　C3200E　　A5判　339頁　3,200円

[国連研究⑦] 国連による国家構築活動での人的側面・信頼醸成活動、あるいは持続可能性の目標および指標などから、持続可能な開発の新しい理論的、実践的な展開過程を描き出す。　　　　　　　　　　　　　(2006.5)

日本国際連合学会編
平和構築と国連
87791-171-3　C3032　　A5判　321頁　3,200円

[国連研究⑧] 包括的な紛争予防、平和構築の重要性が広く認識されている今日、国連平和活動と人道援助活動との矛盾の克服など平和構築活動の現場からの提言を踏まえ、国連による平和と安全の維持を理論的にも追究する。　(2007.6)

日本国際連合学会編
国連憲章体制への挑戦
87791-185-0　C3032　　A5判　305頁　3,200円

[国連研究⑨] とりわけ今世紀に入り、変動著しい世界社会において国連もまた質的変容を迫られている。「国連憲章体制への挑戦」とも言える今日的課題に向け、特集論文、研究ノートなどが理論的追究を展開する。　(2008.6)

日本国際連合学会編
国連研究の課題と展望
87791-195-9　C3032　　A5判　309頁　3,200円

[国連研究⑩] 地球的・人類的課題に取り組み、国際社会で独自に行動する行為主体としての国連行動をたどり未来を展望してきた本シリーズの第10巻目の本書で、改めて国連に関する「研究」に光を当て学問的発展を期す。　(2009.6)

日本国際連合学会編
新たな地球規範と国連
87791-210-9　C3032　　A5判　297頁　3,200円

[国連研究⑪] 新たな局面に入った国連の地球規範；感染症の問題、被害者の視点からの難民問題、保護する責任、企業による人権侵害と平和構築、核なき世界の課題など。人や周囲への思いやりの観点から考える。　　　　(2010.6)

国際法

日本国際連合学会編
安全保障をめぐる地域と国連
87791-220-8　C3032　　　　　　A5判　285頁　3,200円

[国連研究⑫] 人間の安全保障など、これまでの安全保障の再検討が要請され、地域機構、準地域機構と国連の果たす役割が新たに問われている。本書では国際機構論、国際政治学などの立場から貴重な議論が実現した。　　　　　　　　(2011.6)

日本国際連合学会編
日本と国連
――多元的視点からの再考
87791-230-7　C3032　　　　　　A5判　301頁　3,200円

[国連研究⑬] 第13巻目を迎えた本研究は、多元的な視点、多様な学問領域、学会内外の研究者と実務経験者の立場から展開され、本学会が国際的使命を果たすべく「日本と国連」との関係を整理・分析し展望を試みる。　　　　　　　　(2012.6)

日本国際連合学会編
「法の支配」と国際機構
――その過去・現在・未来
87791-250-5　C3032　　　　　　A5判　281頁　3,200円

[国連研究⑭] 国連ならびに国連と接点を有する領域における「法の支配」の創造、執行、監視などの諸活動に関する過去と現在を検証し、「法の支配」が国際機構において持つ現代的意味とその未来を探る。　　　　　　　　(2013.6)

日本国際連合学会編
グローバル・コモンズと国連
87791-260-4　C3032　　　　　　A5判　315頁　3,200円

[国連研究⑮] 公共圏、金融、環境、安全保障の分野から地球公共財・共有資源「グローバル・コモンズ」をさまざまな角度から分析し、国連をはじめとした国際機関の課題および運動の方向を追究する。　　　　　　　　(2014.6)

日本国際連合学会編
ジェンダーと国連
87791-269-7　C3032　　　　　　A5判　301頁　3,200円

[国連研究第⑯] 国連で採択された人権文書、国連と国際社会の動き、「女性・平和・安全保障」の制度化、国連におけるジェンダー主流化と貿易自由化による試み、国連と性的指向・性自認など国連におけるジェンダー課題提起の書。　　(2016.6)

日本国際連合学会編
『国連：戦後70年の歩み、課題、展望』
(『国連研究』第17号)
87791-274-1　C3032　　　　　　A5判　329頁　3,200円

[国連研究⑰] 創設70周年を迎えた国連は第二次世界大戦の惨禍を繰り返さない人類の決意として「平和的生存」の実現を掲げた。しかし絶えない紛争の下、「国連不要論」を乗り越え、いま国連の「課題」および「展望」を追う。　(2016.6)

日本国際連合学会編
多国間主義の展開
87791-283-3　C3032　　　　　　A5判　323頁　3,200円

[国連研究⑱] 米トランプ政権が多国間主義の撤退の動きを強めるなか、諸問題に多くの国がともに解決を目指す多国間主義、国連の活動に日本はどう向き合うのか。若手研究者が歴史的課題に果敢に挑戦する。　　　　　　(2017.6)

望月康恵
人道的干渉の法理論
87791-120-0　C3032　　　　　　A5判　317頁　5,040円

[21世紀国際法学術叢書①] 国際法上の人道的干渉を、①人権諸条約上の人権の保護と人道的干渉における人道性、②内政不干渉原則、③武力行使禁止原則と人道的「干渉」との関係を事例研究で跡づけつつ、具体的かつ実行可能な基準を提示する。　　　　　　　　(2003.3)

吉村祥子
国連非軍事的制裁の法的問題
87791-124-3　C3032　　　　　　A5判　437頁　5,800円

[21世紀国際法学術叢書②] 国際連合が採択した非軍事的制裁措置に関する決議を取り上げ、決議に対する国家による履行の分析、私人である企業に対して適用される際の法的効果を実証的に考察する。　　　　　　　　(2003.9)

国際法

滝澤美佐子
国際人権基準の法的性格
87791-133-2 C3032　　A5判 337頁 5,400円

[21世紀国際法学術叢書③] 国際人権基準の「拘束力」および法的性格の解明を目指す本書は、国際法と国際機構の法秩序とのダイナミズムによって国際人権基準規範の実現が促されていることを明らかにする。　(2004.2)

小尾尚子
難民問題への新しいアプローチ
——アジアの難民本国における難民高等弁務官事務所の活動
87791-134-0 C3032　　A5判 289頁 5,600円

[21世紀国際法学術叢書④] UNHCRのアジアでの活動に焦点を当て、正統性の問題あるいはオペレーション能力の課題を考察し、難民本国における活動が、新しい規範を創りだし、国際社会に定着してゆく過程を描く。　(2004.7)

坂本まゆみ
テロリズム対処システムの再構成
87791-140-5 C3032　　A5判 279頁 5,600円

[21世紀国際法学術叢書⑤] 条約上の対処システム、武力紛争としてのテロリズム対処、テロリズムに対する集団的措置、などを法理論的に整理し、効果的なテロリズムに対する取り組みを実践的に追及する。　(2004.12)

一之瀬高博
国際環境法における通報協議義務
87791-161-8 C3032　　A5判 307頁 5,000円

[21世紀国際法学術叢書⑥] 手続き法としての国際環境損害の未然防止を目的とする通報協議義務の機能と特徴を、事後賠償の実体法としての国際法の限界とを対比・分析することを通して明らかにする。　(2004.2)

石黒一憲
情報通信・知的財産権への国際的視点
906319-13-0 C3032　　A5判 224頁 3,200円

国際貿易における規制緩和と規制強化の中での国際的に自由な情報流通について論ずる。国際・国内両レベルでの標準化作業と知的財産権問題の接点を巡って検討し、自由貿易と公正貿易の相矛盾する方向でのベクトルの本質に迫る。　(1990.4)

廣江健司
アメリカ国際私法の研究
——不法行為準拠法選定に関する方法論と判例法状態
906319-46-7 C3032　　A5判 289頁 4,660円

アメリカ合衆国の抵触法における準拠法選定の方法論を検討する。準拠法選定に関する判例法は、不法行為事件を中心に発展してきているので法域外の要素を含む不法行為を中心に、その方法論を検討し、その判例法状態を検証する。　(1994.3)

廣江健司
国際取引における国際私法
906319-56-4 C1032　　A5判 249頁 3,107円

国際民事訴訟法事件とその国際私法的処理について基礎的な法理論から法実務への架橋となる法情報を提供する。国際取引法の基礎にある法問題、国際私法の財産取引に関する問題、国際民事訴訟法の重要課題を概説した基本書である。　(1995.1)

高橋明弘
知的財産の研究開発過程における競争法理の意義
87791-122-7 C3032　　A5判 361頁 6,200円

コンピュータプログラムのリバース・エンジニアリングを素材に、財産権の社会的側面を、独占（競争制限）、労働のみならず、知的財産並びに環境問題で生じる民法上の不法行為及び権利論の解決へ向けての法概念としても捉える。　(2003.6)

久保田隆
資金決済システムの法的課題
87791-126-× C3032　　A5判 305頁 5,200円

我々に身近なカード決済、ネット決済や日銀ネット、外為円決済システム等、資金決済システムの制度的・法的課題を最新情報に基づき実務・学問の両面から追究した意欲作。金融に携わる実務家・研究者および学生必読の書。　(2003.6)